Praise for **STRANGER THAN WE CAN IMAGINE**
An Indigo Best Book of the Year, 2015

"*Stranger Than We Can Imagine* begins with relativity, then explains modern art, individualism, nihilism, the space race, chaos theory and more. It's brain-bursting stuff, but worth the effort."

—*Metronews.ca*

"A beautiful, erudite, funny and enlightening tour of the widening boundaries of uncertainty revealed in the twentieth century, and who doesn't need a book that explains quantum behaviour with a boxing bout between Putin and a kangaroo?" —Robin Ince

"Hugely entertaining and thought-provoking." —Scott Pack

"An idiosyncratic, always-provocative ʼ would just as soon forget. . . . Full of uɪ written, this is an absorbing tour of th

"Higgs broadens his intellectual reach to encompass modernism, situationism, chaos theory, indeterminacy and almost every other byway of that epoch. . . . A fine example of learning worn lightly."

—*New Scientist*

"Higgs recounts [the twentieth century] with wide-ranging erudition and a delightful deadpan humour." —*Irish Examiner*

"A brisk and surprisingly cohesive cultural history of the 'dark woods' of the 20th century." —*Publishers Weekly*

"Excellent, and consistently intriguing." —*CHOICE* magazine

"A challenging book that stimulates the reader to think radically."

—*Daily Mail*

STRANGER
THAN WE CAN
IMAGINE

By the same author

Non-fiction
*The KLF: Chaos, Magic and the Band Who
Burned a Million Pounds*
I Have America Surrounded: The Life of Timothy Leary
Our Pet Queen: A New Perspective on Monarchy

Fiction
The Brandy of the Damned
The First Church on the Moon

STRANGER
THAN WE CAN
IMAGINE

MAKING SENSE OF THE
TWENTIETH CENTURY

JOHN HIGGS

SIGNAL
McCLELLAND
& STEWART

*For Lia, the twentieth century's post-credits twist,
and for Isaac, the pre-game cutscene of the twenty-first century.
All love, Dad x*

Library and Archives Canada Cataloguing in Publication

Higgs, John, author
Stranger than we can imagine : an alternative history of the 20th century / John Higgs.

Includes bibliographical references and index.
ISBN 978-0-7710-3849-5 (paperback)

1. Civilization, Modern – 20th century. 2. Twentieth century.
I. Title.

CB425.H54 2016   909.82   C2016-900781-2

eBook ISBN 978-0771-03848-8

Typeset in Minion Pro by Input Data Services Ltd, Bridgwater, Somerset

Cover design by Andrew Roberts
Cover image: Salvador Dalí, *Lobster Telephone*, 1936 © Salvador Dalí, Fundació Gala-
Salvador Dalí/ SODRAC (2016)/ Carl de Souza/AFP

Cover design: Adapted from an original by Orionbooks

Printed and bound in the United States of America

McClelland & Stewart,
a division of Random House of Canada Limited
a Penguin Random House Company
www.penguinrandomhouse.ca

1 2 3 4 5   20 19 18 17 16

CONTENTS

'We needed to do what we wanted to do'
Keith Richards

Murdering Airplane *by Max Ernst, 1920* (Bridgeman/© ADAGP Paris & DACS London 2015)

In 2010, the Tate Modern gallery in London staged a retrospective of the work of the French post-impressionist painter Paul Gauguin. To visit this exhibition was to spend hours wandering through Gauguin's vision of a romanticised South Pacific in late nineteenth-century Tahiti. This was a world of vivid colour and guilt-free sexuality. Gauguin's paintings saw no distinction between mankind, divinity and nature, and by the time you reached the end of the exhibition you felt as if you understood Eden.

Visitors were then spat out next to the Tate's twentieth-century gallery. There was nothing to prepare them for how brutal walking out of one and into the other would be.

Here were the works of Picasso, Dalí, Ernst and many others. You immediately wondered if the lighting was different, but it was the art that made this room feel cold. The colour palette was predominantly browns, greys, blues and blacks. Splashes of vivid red appeared in places, but not in ways that comforted. With the exception of a later Picasso portrait, greens and yellows were entirely absent.

These were paintings of alien landscapes, incomprehensible structures and troubled dreams. The few human figures that were present were abstracted, formal, and divorced from contact with the natural world. The sculptures were similarly antagonistic. One example was Man Ray's *Cadeau*, a sculpture of an iron with nails sticking out of its base in order to rip to shreds any fabric you attempted to smooth. Encountering all this in a state of mind attuned to the visions of Gauguin was not recommended. There was no compassion in that room. We had entered the abstract realm of theory and concept. Coming directly from work that spoke to the heart, the sudden shift to work aimed solely at the head was traumatic.

Gauguin's work ran up to his death in 1903, so we might have expected a smoother transition into the early twentieth-century gallery. True, his work was hardly typical of his era and only widely appreciated after his death, but the jarring transition still leaves us struggling to answer a very basic question: what the hell happened, at the beginning of the twentieth century, to the human psyche? The Tate Modern is a suitable place to ask questions like this, as it stands as a kind of shrine to the twentieth century. The meaning of the word 'modern' in the art world means that it will be forever associated with that period. Seen in this light, the popularity of the gallery reveals both our fascination with those years and our desire to understand them.

There was one antechamber which separated the two exhibitions. It was dominated by an outline of a nineteenth-century industrial town by the Italian-Greek artist Jannis Kounellis, drawn directly onto the wall in charcoal. The sketch was sparse and devoid of human figures. Above it hung a dead jackdaw and a hooded crow, stuck to the wall by arrows. I'm not sure what point the artist was trying to make, but for me the room served as a warning about the gallery I was about to enter. It might have been kinder if the Tate had used this room as a form of decompression chamber, something that could prevent the visual art equivalent of the bends.

The dead birds, the accompanying text suggested, 'have been seen as symbolising the death throes of imaginative freedom'. But seen in context between Gauguin and the twentieth century, a different interpretation seemed more appropriate. Whatever it was that had died above that nineteenth-century industrial town, it was not imaginative freedom. On the contrary, that monster was about to emerge from the depths.

Recently I was shopping for Christmas presents and went into my local bookshop for a book by Lucy Worsley, my teenage daughter's favourite historian. If you are lucky enough to have a teenage daughter who has a favourite historian, you don't need much persuading to encourage this interest.

The history books were in the far corner of the fourth floor, at the very top of the building, as if history was the story of crazed ancestors we need to hide in the attic like characters from *Jane Eyre*. The book I wanted wasn't in stock, so I took out my phone to buy it online. I went to shut down an open newspaper app, pressed the wrong icon, and accidentally started a video of a speech made by President Obama a few hours earlier. It was December 2014, and he was talking about whether the hacking of Sony Entertainment, which the President blamed on North Korea, should be regarded as an act of war.

Every now and again there is a moment that brings home how strange life in the twenty-first century can be. There I was in Brighton, England, holding a thin slice of glass and metal which was made in South Korea and ran American software, and which could show me the President of America threatening the Supreme Leader of North Korea. What about this incident would have seemed more incredible at the end of the last century: that this device existed, and allowed me to see the President of the United States while Christmas shopping? That the definition of war could have changed so much that it now included the embarrassing of Sony executives? Or that my fellow shoppers would have been so accepting of my miraculous accidental broadcast?

I was standing next to the twentieth-century history shelves at the time. There were some wonderful books on those shelves, big fat detailed accounts of the century we know most about. Those books act as a roadmap, detailing the journey we took to reach the world we now live in. They tell a clearly defined story of great shifts of geopolitical power: the First World War, the Great Depression, the Second World War, the American Century and the fall of the Berlin Wall. Yet somehow that story fails to lead us into the world we're in now, adrift in a network of constant surveillance, unsustainable competition, tsunamis of trivia and extraordinary opportunity.

Imagine the twentieth century is a landscape, stretching out in front of you. Imagine that the events of its history are mountains, rivers, woods and valleys. Our problem is not that this era is hidden from us, but that we know too much about it. We all know that this landscape contains the mountains of Pearl Harbor, the *Titanic* and South African apartheid. We know that in its centre lies the desolation of fascism and the uncertainty of the Cold War. We know people of this land could be cruel, desperate or living in fear, and we know why. The territory has been thoroughly mapped, catalogued and recorded. It can be overwhelming.

Each of the history books in front of me traces a different path through that territory, but those paths are not as different as you might think. Many are written by politicians or political journalists, or have strong political bias. They take the view that it was politicians that defined these troublesome years, so they follow a path that tells that story. Other books have staked out paths through the art or technology of the period. These are perhaps more useful, but can feel abstract and removed from human lives. And while these paths differ, they do converge along well-trodden highways.

Finding a different path through this territory is daunting. A journey through the twentieth century can seem like an epic quest. The gallant adventurers who embark on it first wrestle with three giants, known by the single names of Einstein, Freud and Joyce. They must pass through the forest of quantum indeterminacy and the castle of conceptual art. They avoid the gorgons of Jean-Paul Sartre and Ayn Rand whose glance can turn them to stone, emotionally if not physically, and they must solve the riddles of the Sphinxes of Carl Jung and Timothy Leary. Then things get difficult. The final challenge is to somehow make it through the swamp of postmodernism. It is not, if we are honest, an appealing journey.

Very few of the adventurers who tackle the twentieth century make it through postmodernism and out the other side. More typically, they admit defeat and retreat to base camp. This is the world as it was understood at the end of the nineteenth century, just over the border, safe in friendly territory. We are comfortable with the great discoveries that emerged up until then. Innovations such as electricity or democracy are comprehensible, and we take them in

6

our stride. But is this really the best place for us? The twenty-first century is not going to make any sense at all seen through nineteenth-century eyes.

The territory of the twentieth century includes dark patches of thick, deep woods. The established paths tend to skirt around these areas, visiting briefly but quickly scurrying on as if fearful of becoming entangled. These are areas such as relativity, cubism, the Somme, quantum mechanics, the id, existentialism, Stalin, psychedelics, chaos mathematics and climate change. They have a reputation for initially appearing difficult, and becoming increasingly bewildering the more they are studied. When they first appeared they were so radical that coming to terms with them meant a major remodelling of how we viewed the world. They seemed frightening in the past, but they don't any more. We're citizens of the twenty-first century now. We made it through yesterday. We're about to encounter tomorrow. We can take the dark woods of the twentieth century in our stride.

So this is our plan: we're going to take a journey through the twentieth century in which we step off the main highways and strike out towards the dark woods. We're aware that a century is an arbitrary time period. Historians talk about the long nineteenth century (1789–1914) or the short twentieth century (1914–91), because these periods contain clear beginnings and endings. But for our purposes 'the twentieth century' will do fine, because we're taking a journey from when things stopped making sense to where we are now.

If we're going to make it through, we're going to have to be selective. There are millions of subjects worthy of inclusion in an account of this period, but we're not going to get very far if we revisit all of our favourites for the sake of nostalgia. There's a wealth of fascinating literature and debate behind everything we find, which we will have to ruthlessly avoid getting bogged down in. We're on a mission, not a cruise. We set out not as historians but as curious travellers, or as adventurers with an agenda, because we are embarking on our travels with a clear sense of what we will be paying attention to.

Our plan is to look at what was genuinely new, unexpected and radical. We're not concerned by the fallout from those ideas, so take it as read that everywhere we visit caused scandal, anger and furious denouncements by the status quo. Those aftershocks are an important part of history, but focusing on them can disguise an emerging pattern. It is the direction that these new ideas were pointing in that we'll pay attention to. They point in a broadly coherent direction.

There's a moment for every generation when memory turns into history. The twentieth century is receding into the distance, and coming into perspective. The events of that century now feel like they belong in the category of history, so this is the right time to take stock.

Here, then, is an alternative route through the landscape of the last century. Its purpose is the same as all paths. It will take you to where you are going.

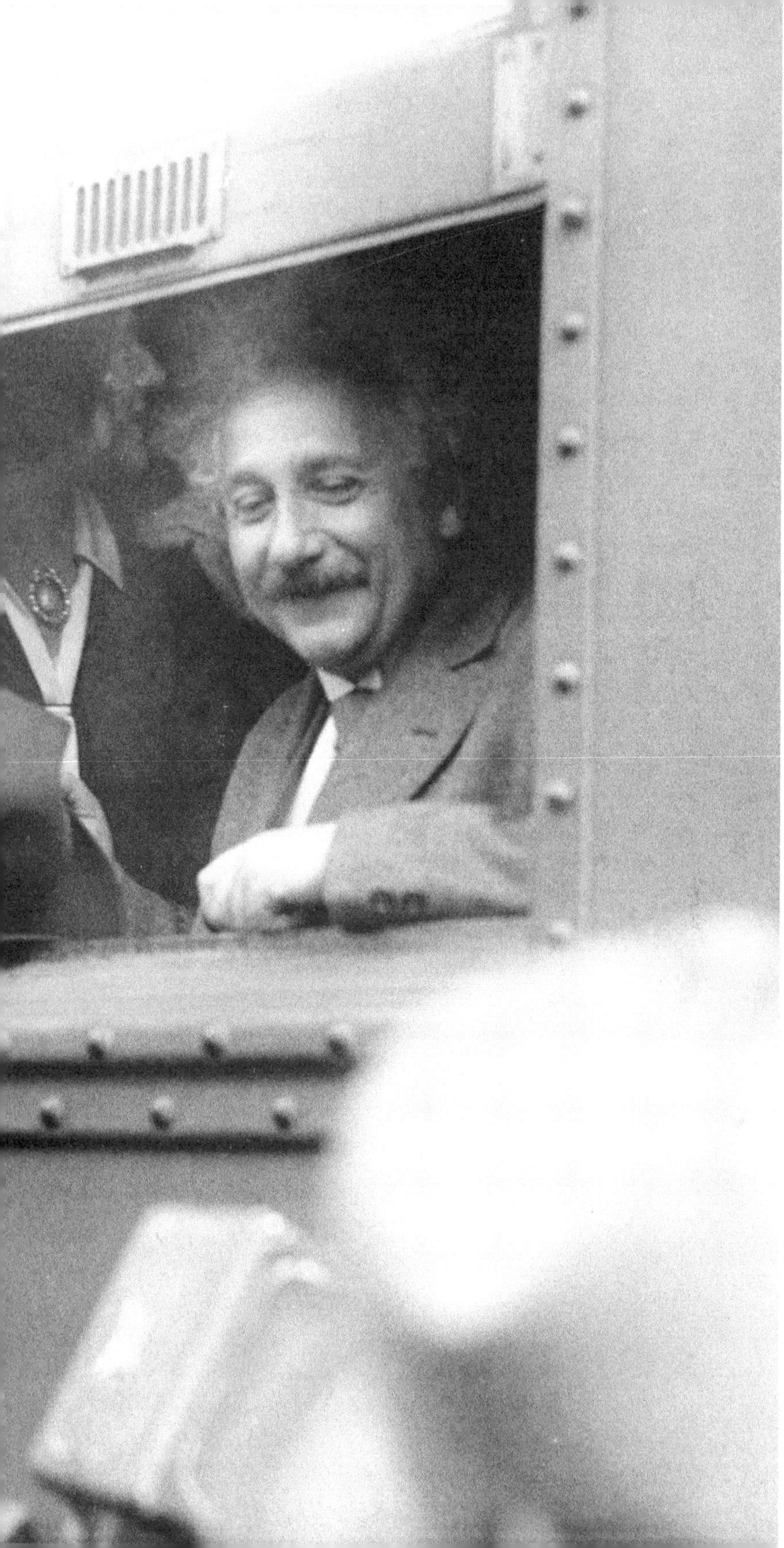

ONE: RELATIVITY

Albert Einstein in Chicago, c.1930 (Transcendental Graphics/Getty)

Deleting the omphalos

On the afternoon of 15 February 1894 the French anarchist Martial Bourdin left his rented room in Fitzroy Street in London. He was carrying a homemade bomb and a large amount of money. It was dry and sunny, and he boarded an open-top horse-drawn tram at Westminster. This took him across the river and on to Greenwich.

After leaving the tram he walked across Greenwich Park towards the Royal Observatory. His bomb exploded early, while he was still in the parkland. The explosion destroyed his left hand and a good chunk of his stomach, but did no damage to the observatory. A group of schoolchildren found him lying on the ground, confused and asking to be taken home. Blood and bodily remains were later found over sixty yards away. Bourdin died thirty minutes after the bomb exploded, leaving no explanation for his actions.

The Polish writer Joseph Conrad's *The Secret Agent* (1907) was inspired by these events. Conrad summed up the general bewilderment about Bourdin's actions when he described the bombing as 'a blood-stained inanity of so fatuous a kind that it is impossible to fathom its origin by any reasonable or even unreasonable process of thought [. . .] One remained faced by the fact of a man blown to pieces for nothing even most remotely resembling an idea, anarchistic or other.'

It wasn't Bourdin's politics that puzzled Conrad. The meaning of the term 'anarchism' has shifted over the last century, so that it is now commonly understood as an absence of rules where everyone can do whatever they like. Anarchism in Bourdin's era was focused more on rejecting political structures than on demands for unfettered personal liberty. Nineteenth-century anarchists weren't claiming the right to total freedom, but they were claiming the right not to be controlled. They recognised, in the words of one of their slogans, 'No gods, no masters'. In terms of Christian theology, they were committing the sin of pride. This was Satan's rebellion and the reason he was cast down from Heaven: *non serviam*, 'I Will Not Serve.'

Nor was Conrad confused by Bourdin's desire to plant a bomb. It was the middle of a violent period of anarchist bombings, which began with the assassination of the Russian tsar Alexander II in 1881 and lasted until the outbreak of the First World War. This was fuelled by the ready availability of dynamite and an anarchist concept called the 'propaganda of the deed', which argued that individual acts of violence were valuable in themselves because they served to inspire others. The anarchist Leon Czolgosz, to give one example, successfully assassinated the President of the United States William McKinley in September 1901.

No, the baffling question was this: if you were an anarchist on the loose in London with a bomb, why would you head for the Royal Observatory at Greenwich? What did it offer as a target that, for example, Buckingham Palace or the Houses of Parliament lacked? Both of these buildings were closer to where Bourdin lived, had a higher profile, and symbolised the power of the state. Why didn't he try to blow those up? It seemed that he had recognised some aspect or quality of the Royal Observatory that he felt was significant enough for him to risk his life to destroy.

In events and stories inspired by the Greenwich bombing, little attention is paid to the target. The explosion was fictionalised in Conrad's novel and that book influenced the American terrorist Ted Kaczynski, better known as the Unabomber. Alfred Hitchcock adapted the story in his 1936 film, *Sabotage*, in which he updated the bomber's journey across London from a horse-drawn tram to a more modern bus. Hitchcock had his bomb explode early when the bus was on The Strand, a spooky fictional precursor to an incident sixty years later when an IRA terrorist accidentally blew himself up on a bus just off The Strand.

But while the target of the bombing may have made little sense to Conrad, that does not mean that it was equally meaningless to Bourdin. As the American cyberpunk author William Gibson would note, 'The future is already here. It is just not very evenly distributed.' Ideas spread unevenly and travel at unpredictable speeds. Perhaps Bourdin glimpsed something remotely resembling an idea that was otherwise invisible to Conrad. As the twentieth century began, the logic behind his target slowly came into focus.

The earth hurtled through the heavens. On its surface, gentlemen checked their pocket watches.

It was 31 December 1900. The earth swung around the sun and minute hands moved around clock faces. When both hands pointed up to twelve it meant that the earth, after travelling thousands of miles, had reached the required position on its yearly circuit. At that moment the twentieth century would begin.

In ancient history there is a concept called an omphalos. An omphalos is the centre of the world or, more accurately, what was culturally thought to be the centre of the world. Seen in a religious context, the omphalos was also the link between heaven and earth. It was sometimes called the navel of the world or the *axis mundi*, the world pillar, and it was represented physically by an object such as a pillar or a stone.

An omphalos is a universal symbol common to almost all cultures, but with different locations. To the ancient Japanese, it was Mount Fuji. To the Sioux, it was the Black Hills. In Greek myth, Zeus released two eagles in order to find the centre of the world. They collided above Delphi, so this became the Greek omphalos. Rome itself was the Roman omphalos, for all roads led there, and later still Christian maps became centred on Jerusalem.

On New Year's Eve 1900, the global omphalos was the Royal Observatory in Greenwich, South London.

The Royal Observatory is an elegant building, founded by Charles II in 1675 and initially designed by Sir Christopher Wren. In 1900 the world was measured from a line that ran north–south through this building. This international standard had been agreed at a conference in Washington DC sixteen years earlier, when delegates from twenty-five countries voted to accept Greenwich as the prime meridian. San Domingo voted against and France and Brazil abstained, but the meeting was largely a formality; 72 per cent of the world's shipping used sea charts that listed Greenwich as zero degrees latitude, and the USA had already based its time zones on Greenwich.

Here, then, was the centre of the world, a seat of science bestowed by royal patronage. It overlooked the Thames in London, the capital city of the largest empire in history. The twentieth century only began when the clocks in this building declared that it had begun, because the calibrations of those clocks were based on the positions of the stars directly above. This modern, scientific omphalos had not lost the link between heaven and earth.

When you visit the observatory today, at dusk or night, you will see the prime meridian represented by a green laser beam, straight and steady, cutting across the sky. It begins at the observatory and is perfectly aligned to zero degrees latitude. The laser did not exist in 1900, of course. The line then was an idea, a mental projection applied to the real world. From here, a net of similar longitude lines stretched outwards to the west and east, reaching further and further around the curve of the globe until they met at the other side. They crossed a similar set of latitude lines, based at the equator, which stretched out to the north and the south. Together this mental web created a universal time zone and positioning system which could synchronise everyone and everywhere on the planet.

On New Year's Eve 1900, people took to the streets in different cities and nations around the world and welcomed in the new century. Nearly a hundred years later, the celebrations that marked the next millennium took place on New Year's Eve 1999 rather than 2000. This was a year early and technically wrong, but few people cared. When the staff at the Greenwich Observatory explained that the twenty-first century didn't actually start until 1 January 2001, they were dismissed as pedants. Yet at the start of the twentieth century the observatory had authority, and the world celebrated as they dictated. Greenwich was the place that mattered. So it was with some satisfaction that the members of Victorian society present checked their watches, awaited the correct time, and witnessed the birth of a new era.

It appeared, on the surface, to be an ordered, structured era. The Victorian worldview was supported by four pillars: Monarchy, Church, Empire and Newton.

The pillars seemed solid. The British Empire would, in a few years, cover a quarter of the globe. Despite the humiliation of the Boer War, not many realised how badly the Empire had been wounded and fewer still recognised how soon it would collapse. The position of the Church looked similarly secure, despite the advances of science. The authority of the Bible may have been contradicted by Darwin and advances in geology, but society did not deem it polite to dwell too heavily on such matters. The laws of Newton had been thoroughly tested and the ordered, clockwork universe they described seemed incontrovertible. True, there were a few oddities that science puzzled over. The orbit of Mercury, for instance, was proving to be slightly different to what was expected. And then there was also the issue of the aether.

The aether was a theoretical substance that could be described as the fabric of the universe. It was widely accepted that it must exist. Experiments had shown time and time again that light travelled in a wave. A light wave needs something to travel through, just as an ocean wave needs water and a sound wave needs air. The light waves that travel through space from the sun to the earth must pass through something, and that something would be the aether. The problem was that experiments designed to reveal the aether kept failing to find it. Still, this was not considered a serious setback. What was needed was further work and cleverer experiments. The expectation of the discovery of the aether was similar to that surrounding the Higgs boson in the days before the CERN Large

Hadron Collider. Scientific wisdom insisted that it must exist, so it was worth creating more and more expensive experiments to locate it.

Scientists had an air of confidence as the new century began. They had a solid framework of knowledge which would withstand further additions and embellishments. As Lord Kelvin was reputed to have remarked in a 1900 lecture, 'there is nothing new to be discovered in physics now. All that remains is more and more precise measurement.' Such views were reasonably common. 'The more important fundamental laws and facts of physical science have all been discovered,' wrote the German-American physicist Albert Michelson in 1903, 'and these are now so firmly established that the possibility of their ever being supplanted in consequence of new discoveries is exceedingly remote.' The astronomer Simon Newcomb is said to have claimed in 1888 that we were 'probably nearing the limit of all we can know about astronomy'.

The great German physicist Max Planck had been advised by his lecturer, the marvellously named Philipp von Jolly, not to pursue the study of physics because 'almost everything is already discovered, and all that remains is to fill a few unimportant holes.' Planck replied that he had no wish to discover new things, only to understand the known fundamentals of the field better. Perhaps unaware of the old maxim that if you want to make God laugh you tell him your plans, he went on to become a founding father of quantum physics.

Scientists did expect some new discoveries. Maxwell's work on the electromagnetic spectrum suggested that there were new forms of energy to be found at either end of his scale, but these new energies were still expected to obey his equations. Mendeleev's periodic table hinted that there were new forms of matter out there somewhere, just waiting to be found and named, but it also promised that these new substances would fit neatly into the periodic table and obey its patterns. Both Pasteur's germ theories and Darwin's theory of evolution pointed to the existence of unknown forms of life, but also offered to categorise them when they were found. The scientific discoveries to come, in other words, would be wonderful but not surprising. The body of knowledge of the twentieth century would be like that of the nineteenth, but padded out further.

Between 1895 and 1901 H.G. Wells wrote a string of books including *The Time Machine, War of the Worlds, The Invisible Man* and *The First Men in the Moon.* In doing so he laid down the blueprints for science fiction, a new genre of ideas and technological speculation which the twentieth century would take to its heart. In 1901 he wrote *Anticipations: An Experiment in Prophecy*, a series of articles which attempted to predict the coming years and which served to cement his reputation as the leading futurist of the age. Looking at these essays with the benefit of hindsight, and awkwardly skipping past the extreme racism of certain sections, we see that he was successful in an impressive number of predictions. Wells predicted flying machines, and wars fought in the air. He foresaw trains and cars resulting in populations shifting from the cities to the suburbs. He predicted fascist dictatorships, a world war around 1940, and the European Union. He even predicted greater sexual freedom for men and women, a prophecy that he did his best to confirm by embarking on a great number of extramarital affairs.

But there was a lot that Wells wasn't able to predict: relativity, nuclear weapons, quantum mechanics, microchips, black holes, postmodernism and so forth. These weren't so much unforeseen, as unforeseeable. His predictions had much in common with the expectations of the scientific world, in that he extrapolated from what was then known. In the words commonly assigned to the English astrophysicist Sir Arthur Eddington, the universe would prove to be not just stranger than we imagine but, 'stranger than we can imagine'.

These unforeseeable new discoveries would not happen in Greenwich or Britain, where the assembled dignitaries were comfortable with the structure of the world. Nor would they appear in the United States, or at least not initially, even though the opening up of the Texas oilfields around this time would have a massive impact on the world to come. At the beginning of the twentieth century it was in the cafés, universities and journals of Germany and the German-speaking people of Switzerland and Austria that the real interest in testing and debating radical ideas lay.

If we had to choose one town as the birthplace of the twentieth century then our prime contender must be Zurich, an ancient city which straddles the River Limmat just north of the Swiss Alps. In the year 1900 it was a thriving town of tree-lined streets and buildings which managed to be both imposing and pretty at the same time. It was here, at the Zurich Polytechnic, that twenty-one-year-old Albert Einstein and his girlfriend Mileva Marić were about to come bottom in their class.

Einstein's career did not then appear promising. He was a rebellious and free-spirited young man who had already renounced both his Jewish religion and his German citizenship. Six months earlier, in July 1899, he clumsily caused an explosion in the physics lab which damaged his right hand and temporarily stopped him from playing his beloved violin. His Bohemian personality caused him to clash with the academic authorities and prevented him from gaining a job as a physicist when he finally graduated. There was little sign that the world of science would take any notice of this stubborn, belligerent young man.

There's been some debate about the role of Marić, whom he married in 1903, in Einstein's early achievements. Marić was not the sort of woman that early twentieth-century society approved of. She was one of the first women in Europe to study mathematics and physics. There was a good deal of prejudice about her Slavic background, and the fact that she suffered from a limp. Einstein, however, had no interest in the dull prejudices of his time. There was an intensity about her that entranced him. She was, as his many love letters make clear, his 'little witch' and his 'wild street urchin' and, for a few years at least, they were everything to each other.

Marić believed in Einstein. A muse can bring out the genius inside a scientist just as with an artist. It took a rare and youthful arrogance to even consider attempting what Einstein was about to do. With the love of Marić validating his belief in himself, and the intellectual freedom he never would have had if he'd found an academic position, Albert Einstein rewrote our understanding of the universe.

'So what are you up to,' Einstein wrote to his friend Conrad Habicht in May 1905, 'you frozen whale, you smoked, dried piece of soul? Such a solemn air of silence has descended between us that I almost feel as if I am committing a sacrilege with some inconsequential babble...'

During the 'inconsequential babble' of the letter that followed Einstein casually described four papers that he was working on. Any one of them would have been a career-making achievement. That he produced all four in such a short space of time is almost unbelievable. Science historians have taken to referring to 1905 as Einstein's 'miracle year'. It is not often that historians of science reach for the word 'miracle'.

Einstein's work in 1905 recalls the achievements of Isaac Newton in 1666, when the plague closed Cambridge University and Newton returned to his mother's home in rural Lincolnshire. He used the time to develop calculus, a theory of colour and the laws of gravity, immortalising himself as Britain's greatest scientific genius as he did so. Einstein's achievement is more impressive when you consider that he wasn't idling about under apple trees but holding down a full-time job. He was then employed at the patent office in Bern, having failed to gain employment as a physicist. Incredibly, he wrote these four papers in his spare time.

'The first [of his proposed papers] deals with radiation and the energy properties of light and is very revolutionary,' he wrote. This is no overstatement. In it he argued that light consists of discrete units, or what we now call photons, and that the aether doesn't exist. As we shall see later, this paper inadvertently laid the groundwork for quantum physics and a model of the universe so strange and counterintuitive that Einstein himself would spend most of his life trying to deny the implications.

'The second paper is a determination of the true sizes of atoms.' This was the least controversial of the papers, being useful physics that did not overturn any established ideas. It gained Einstein his doctorate. His third paper used statistical analysis of the movement of visible particles in water to prove beyond doubt the existence of atoms, something that had been widely suspected but never conclusively proved.

Einstein's most significant discovery came from pondering a seeming contradiction between two different laws of physics. 'The fourth paper is only a rough draft at this point, and is an electrodynamics of moving bodies which employs a modification of the theory of space and time,' he wrote. This would become the Special Theory of Relativity. Together with the broader General Theory of Relativity he produced ten years later, it overturned the graceful, clockwork universe described by Newton.

Relativity showed that we lived in a stranger, more complex universe where space and time were no longer fixed, but could be stretched by mass and motion. This was a universe of black holes and warped space-time that seemed to have little in common with the everyday world in which we live. Relativity is often presented in ways that make it appear incomprehensible, but the core idea at its heart can be grasped surprisingly easily.

Imagine the deepest, darkest, emptiest chunk of space possible, far removed from stars, planets or any other influence. In this deep void imagine that you are floating, snug and warm in a space suit. Importantly, imagine that you are not moving.

Then imagine that a cup of tea comes slowly floating past, and eventually disappears into the distance.

At first glance, this scenario sounds reasonable. Newton's First Law says that an object will continue to remain at rest, or will move in a straight line at a constant velocity, unless some external force acts on it. Clearly, this is a perfect description of the behaviour of both you and the cup of tea.

But how could we say that you were at rest? Einstein would ask. How do we know that it's not the cup of tea that's at rest, and you that are moving past it? Both situations would appear identical from your point of view. And also, from the point of view of the cup of tea.

Galileo was told in the 1630s that it wasn't possible the earth was going around the sun, because we on earth do not feel like we are moving. But Galileo knew that if you were moving smoothly, without accelerating or decelerating, and if there were no visible or audible clues to movement, then you would not be aware of your motion. He argued that you cannot claim to be 'at rest', because it is impossible to tell the difference between a moving object and a stationary one without some form of external reference to compare it against.

This may sound like a dubious, pedantic point. Surely, you might think, you are either moving or not moving, even if there's nothing else around. How could anyone claim that the statement 'you are at rest' is absurd or meaningless?

Schoolchildren are taught to plot the position of objects by drawing diagrams that show their distance from a fixed point in terms of height, length and depth. These are called the x, y and z axis, and the fixed point is usually called O or the origin. This is an omphalos, from which all the other distances are measured. The territory marked out by these x, y and z axes is called Cartesian space. In this framework, it would be simple to tell whether the astronaut and the cup of tea were static or moving by noting whether their coordinates in Cartesian space changed over time.

But if you had shown that illustration to Einstein he would have leant over with an eraser and removed the origin, and then rubbed out the x, y and z axis while he was at it.

He wouldn't be deleting 'space' itself. He would be removing the frame of reference that we were using to define space. He would do this because it was not a feature of the real world. That framework of Cartesian space is a product of our minds, like the longitude lines stretching away from Greenwich, which we project onto the cosmos in order to get a grip on it. It does not really exist. Also, it is arbitrary. That framework could have been centred anywhere.

Instinctively we feel that we or the tea must be moving – or not – against some form of definitive 'background'. But if there is a definitive background, what could it be?

In our everyday lives the solid ground beneath our feet is a point of reference that we unconsciously judge everything by. Living with such a clear fixed point makes it hard to imagine one not existing. But how fixed is the ground? We have known that continents are slowly moving since the acceptance of plate tectonics in the 1960s. If we are seeking a fixed point, it is not the land that we stand on.

Could we instead define our position with the very centre of the earth? This isn't fixed either, because the earth is moving around the sun at over 100,000 km/h. Or perhaps we can define the sun as our fixed point? The sun is moving at 220 km/s around the centre of the Milky Way galaxy. The Milky Way, in turn is moving at 552 km/s, relative to the rest of the universe.

What of the universe itself? As a last-ditch and somewhat extreme attempt to locate a fixed point, could we not declare the centre of the universe to be our omphalos? The answer, once more, is no. There is no 'centre of the universe', as we will see later, but for now we can also reject the idea for being ridiculously impractical.

So how can we say anything definite about our position, or that of the cup of tea? There may not be a real 'fixed point' which we can use, but we are still free to project our own frames of reference wherever we like. We can create a reference frame centred on ourselves, for example, which allows us to say that the tea is moving relative to us. Or we can create one centred on the tea, which would mean that we were moving relative to the cup. What we can't do is say that one of these frames of reference is correct or more valid than the other. To say that the tea moved past us would be to declare our innate, tea-ist prejudice.

There is an apt example of how one frame of reference is no more valid than another in Einstein's 1917 book, *Relativity*. In the original German-language edition, he used Potsdamer Platz in Berlin as the frame of reference in one example. When the book was translated into English, this was changed to Trafalgar Square in London. By the time the book was out of copyright and made available online as an eBook, this had been changed to Times Square in New York because, in the opinion of the editor, 'this is the most well known/ identifiable location to English speakers in the present day.' What is important about the reference point, in other words, is that it has been defined as the reference point. Practically, it could be anywhere.

The first step towards understanding relativity, then, is to accept this: a statement of position is only meaningful when it has been defined along with its frame of reference. We can choose whatever frame of reference we like, but we can't say that it has more validity than any other.

With that in mind, let us return to Zurich in 1914.

Einstein gets on a steam train in Zurich and travels to Berlin. He is leaving his wife Marić and their two surviving children in order to begin a new life with his cousin, who will later become his second wife. Imagine that the train travels in a straight line at a constant speed of 100 km/h, and that at one point he stands, holds a sausage at head height, and drops it on the floor.

This raises two questions: how far does the sausage fall, and why is he leaving his wife? Of these two questions Einstein would have found the first one to be the most interesting, so this is what we will focus on.

Let us say he holds the sausage up to a height of five feet above the train floor and drops it. It lands, as you would expect, near to his scuffed shoes, directly below his raised hand. We can say that it has fallen five feet exactly. As we have just seen, such a statement only makes sense when the frame of reference is defined. Here we take Einstein's frame of reference, that of the inside of the train carriage, and we can say that relative to that, the sausage fell five feet.

What other frames of reference could we use? Imagine there is a mouse on the railway track, and that the train rumbles safely over the mouse as Einstein drops his sausage. How far would the sausage fall if we use this mouse as a point of reference?

The sausage still starts in Einstein's hand and lands by his feet. But, as far as the mouse is concerned, Einstein and the sausage are also moving over him during the sausage's fall. During the period between Einstein letting go of the sausage and it hitting the floor, it will have moved a certain distance along the track. The position of his feet when the sausage lands is further down the track than the position of the hand at the moment it was dropped. The sausage has still fallen five feet downwards, relative to the mouse, but it has also travelled a certain distance in the direction the train is travelling in. If you were to measure the distance taken by the sausage between the hand and the floor, relative to the mouse, its path would be at an angle rather than pointing straight down, and that means it would have travelled further than five feet.

This is, instinctively, something of a shock. The distance that the sausage moves changes when it is measured from different frames of reference. The sausage travelled further from the mouse's perspective than it did from Einstein's. And, as we have seen, we cannot say that one frame of reference is more valid than any other. If this is the case, how can we make any definitive statements about distance? All we can do is say that the sausage fell a certain distance relative to a particular frame of reference, and that distance can be different when measured from other frames of reference.

This is only the beginning of our troubles. How long did the sausage take to fall? As you can appreciate, a sausage that falls more than five feet will take longer than one that just falls five feet. This leaves us with the slightly disturbing conclusion that dropping the sausage took less time from Einstein's point of view than it did for the mouse.

No!

Just as we live with the constant fixed point of the ground beneath us, we believe that there is a constant, universal time ticking away in the background. Imagine the bustle of commuters crossing Westminster Bridge in London, with the Houses of Parliament and the clock face of Big Ben up above them. The clock is suspended above the suited people below, ticking away with absolute regularity, unaffected by the lives going on beneath it. This is similar to how we intuitively feel time must work. It is beyond us, and unaffected by what we do. But Einstein realised that this was not the case. Time, like space, differs according to circumstances.

All this seems to leave us in a tricky situation. Measurements of time and space differ depending on which frame of reference we use, but there is no 'correct' or 'absolute' frame of reference that we can rely on. What is observed is dependent, in part, on the observer. At first glance, it appears that this leaves us in a desperate situation, one where every measurement is relative and cannot be said to be definitive or 'true'.

In order to escape from this hole, Einstein reached for mathematics.

According to well-established physics, light (and all other forms of electromagnetic radiation) must always move at a particular speed when it travels through a vacuum. This speed, nearly 300,000,000 metres per second, is known to mathematicians as 'c' and to non-mathematicians as 'the speed of light'. How, though, can this be the case when, as we have seen, measurements differ depending on the frame of reference?

In particular, there is the law of addition of velocities to consider. Consider a scene in a James Bond movie where James Bond is shot at by the henchman of an evil villain. We don't need to worry whether Bond will be killed, as henchmen are notoriously bad shots. Instead, let us worry about how fast that bullet will be travelling when it sails harmlessly over his head. Imagine, for the sake of argument, the speed of the bullet from the gun was 1,000 mph. If the henchman was driving towards Bond on a snowmobile when he fired, and if the snowmobile was travelling at 80 mph, then the velocity of the bullet would be these speeds added together, or 1,080 mph. If Bond was skiing away at 20 mph at the time, then this would also need to be factored in, and the bullet would then have a velocity relative to Bond of 1,060 mph.

Now back to Einstein on the steam train, who has swapped his sausage for a torch which he shines along the length of the dining carriage. From his point of view, the photons emitted from the torch travel at the speed of light (strictly speaking the train would need to be in a vacuum for them to reach this speed, but we'll ignore such details so that he doesn't suffocate). Yet for a static observer who was not on the train, such as the mouse from earlier or a badger underneath a nearby tree, the photons would appear to travel at the speed of light plus the speed of the train, which, clearly, is a different speed to the speed of light. Here we have what appears to be a fundamental contradiction in the laws of physics, between the law of addition of velocities and the rule that electromagnetic waves must always travel at the speed of light.

Something is not right here. In an effort to resolve this difficulty we might ask if the law of addition of velocities is in some way flawed, or if the speed of light is as certain as claimed. Einstein looked at these two laws, decided that they were both fine, and came to a startling conclusion. The speed of light, nearly 300,000,000 metres per second, was not the problem. It was the 'metres' and 'seconds' that were the problem. Einstein realised that when an object travelled at speed, space got shorter and time moved slower.

Einstein backed up this bold insight by diving into the world of mathematics. The main tool that he used was a technique called a Lorentz transformation, which was a method that allowed him to convert between measurements taken from different frames of reference. By mathematically factoring out those different reference frames, Einstein was able to talk objectively about time and space and demonstrate exactly how they were affected by motion.

Just to complicate matters further, it is not just motion that shrinks time and space. Gravity has a similar effect, as Einstein discovered in his General Theory of Relativity ten years later. Someone living in a ground-floor flat will age slower than their neighbour living on the first floor because the gravity is fractionally stronger closer to the ground. The effect is tiny, of course. The difference would be less than a millionth of a second over a lifespan of eighty years. Yet it is a real effect nonetheless, and it has been measured in the real world. If you get two identical, highly accurate clocks and put one on an aeroplane while keeping the other still, the clock that has flown at speed will show that less time has passed than that measured by the static clock. The satellites that your car's satellite navigation system rely on are only accurate because they factor in the effect of the earth's gravity and their speed when they calculate positions. It is Einstein's maths, not our common-sense concept of three-dimensional space, which accurately describes the universe we live in.

How can non-mathematicians understand Einstein's mathematical world, which he called *space-time*? We are trapped in the reference frames that we use to understand our regular world, and we are unable to escape to his higher mathematical perspective where their contradictions melt away. Our best hope is to look downwards at a more constrained perspective that we can understand, and use that as an analogy for imagining space-time.

Imagine a flat, two-dimensional world where there is length and breadth but no height. The Victorian teacher Edwin Abbott Abbott wrote a wonderful novella about such a place, which he called *Flatland*. Even if you are not familiar with this book, you can picture such a world easily by holding a piece of paper in front of you and imagining that things lived in it.

If this piece of paper were a world populated by little flat beings, as in Abbott's story, they would not be aware of you holding the paper. They could not comprehend our three-dimensional world, having no concept of 'up'. If you were to bend and flex the paper they would not notice, for they have no understanding of the dimension in which these changes are taking place. It would all seem reassuringly flat to them.

Now imagine that you roll the paper into a tube. Our little flat friends will still not realise anything has happened. But they will be surprised when they discover that, if they walk in one direction for long enough, they no longer reach the end of the world but instead arrive back where they started. If their two-dimensional world is shaped like a tube or a globe, like the skin of a football, how could these people explain those bewildering journeys that do not end? It took mankind long enough to accept that we live on a round planet even though we possessed footballs and had the advantage of understanding the concept of globes, yet these flat critters don't even have the *idea* of globes to give them a clue. They will need to wait until there comes among them a flat equivalent of Einstein, who would use strange arcane mathematics to argue that their flat world must exist in a higher-dimensional universe, where some three-dimensional swine was bending the flat world for their own unknowable reasons. The other flat critters would find all this bewildering, but given time they will discover that their measurements, experiments and regular long walks fit Flat Einstein's predictions. They would then be confronted with the realisation that there is a higher dimension after all, regardless of how ludicrous this might seem or how impossible it is to imagine.

We are in a similar position to these flat creatures. We have measurements and data that can only be explained by the mathematics of space-time, yet space-time remains incomprehensible to the majority of us. This is not helped by the glee with which scientists describe the stranger aspects of relativity instead of explaining what it is and how it relates to the world we know. Most people will have heard the example of how, should a distant observer see you fall into a black hole, you would appear to take an infinite amount of time to fall even though you yourself thought you fell quickly. Physicists love that sort of stuff. Befuddlement thrills them, but not everyone benefits from being befuddled.

It is true that space-time is a deeply weird place from a human perspective, where time behaves like any other dimension and concepts such as 'future' and 'past' do not apply as we normally understand them. But the beauty of space-time is that, once understood, it removes strangeness, not creates it. All sorts of anomalous measurements, such as the orbit of Mercury or the way light bends around massive stars, lose their mystery and contradictions. The incident where the cup of tea may or may not pass you in deep space becomes perfectly clear and uncontroversial. Nothing is at rest, unless it is defined as being so.

General Relativity made Einstein a global celebrity. He made an immediate impression on the public, thanks to press photographs of his unkempt hair, crumpled clothes and kind, smiling eyes. The idea of a 'funny little man' from the European continent with a mind that could see what others could not was a likeable archetype, one which Agatha Christie put to good use when she created Poirot in 1920. The fact that Einstein was a German Jew only added to the interest.

The reception of Einstein and relativity shows a world more interested in the man than his ideas. Many writers took an almost gleeful pleasure in their failure to understand his theories, and the idea that relativity was impossible for normal people to comprehend soon took hold. Contemporary press reports claimed that there were only twelve people in the world who could understand it. When Einstein visited Washington in 1921 the Senate felt the need to debate his theory, with a number of Senators arguing that it was incomprehensible. President Harding was happy to admit that he didn't understand it. Chaim Weizmann, later the first President of Israel, accompanied Einstein on an Atlantic crossing. 'During the crossing Einstein explained relativity to me every day,' he remarked, 'and by the time we arrived I was fully convinced that he really understands it.'

Relativity arrived too late for the anarchist Martial Bourdin. He wanted to destroy Greenwich Observatory, which was symbolically the omphalos of the British Empire and its system of order which stretched around the entire globe. But omphaloi, Albert Einstein taught us, are entirely arbitrary. Had Bourdin waited for the General Theory of Relativity, he might have realised that it was not necessary to build a bomb. All that was needed was to recognise that an omphalos was a fiction in the first place.

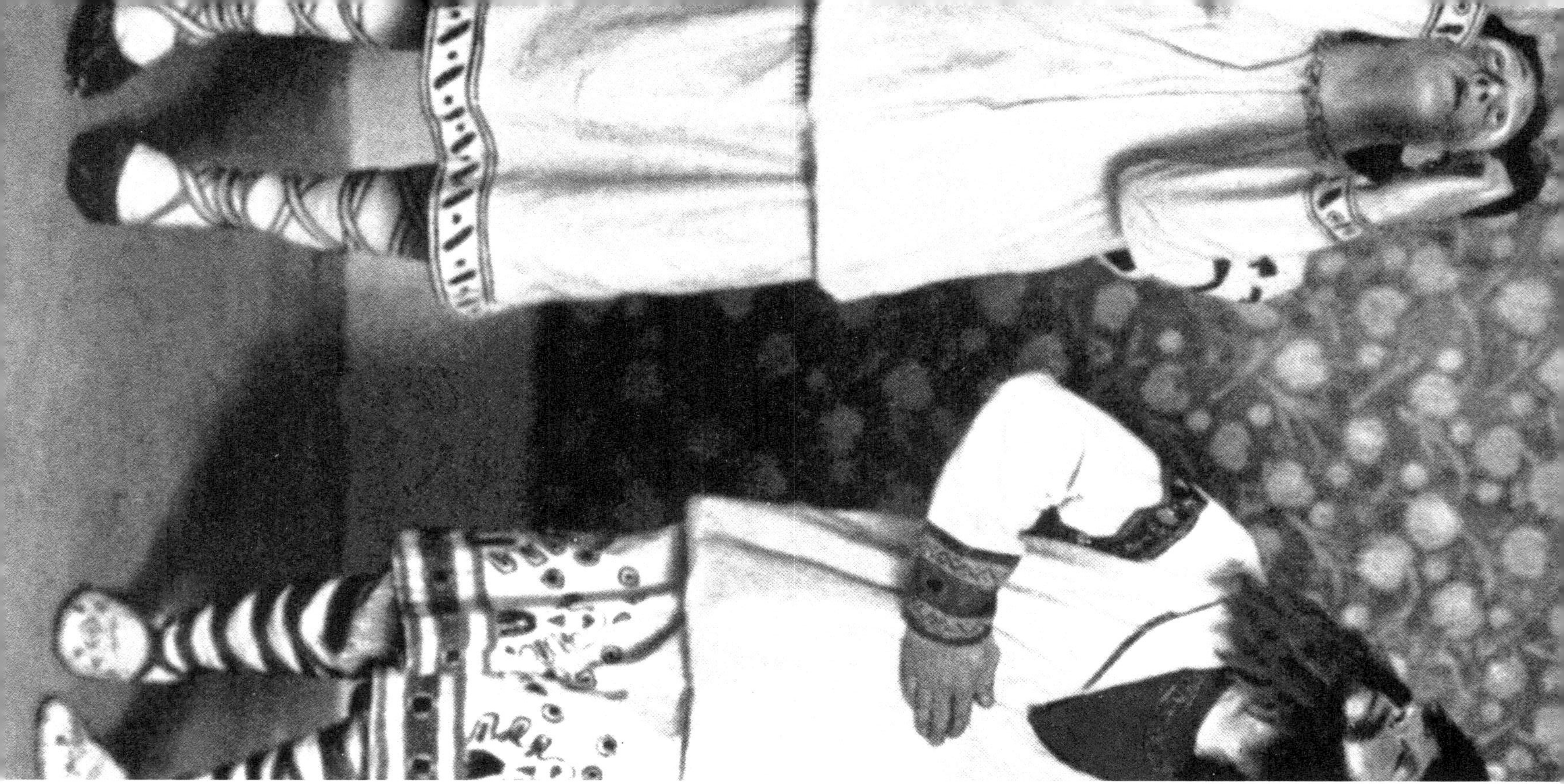

TWO: MODERNISM

A scene from The Rite of Spring *at the Théâtre des Champs-Élysées, 1913* (Keystone-France/Getty)

The shock of the new

In March 1917 the Philadelphia-based modernist painter George Biddle hired a forty-two-year-old German woman as a model. She visited him in his studio, and Biddle told her that he wished to see her naked. The model threw open her scarlet raincoat. Underneath, she was nude apart from a bra made from two tomato cans and green string, and a small birdcage housing a sorry-looking canary, which hung around her neck. Her only other items of clothing were a large number of curtain rings, recently stolen from Wanamaker's department store, which covered one arm, and a hat which was decorated with carrots, beets and other vegetables.

Poor George Biddle. There he was, thinking that he was the artist and that the woman in front of him, Baroness Elsa von Freytag-Loringhoven, was his model. With one quick reveal the baroness announced that she was the artist, and he simply her audience.

Then a well-known figure on the New York avant garde art scene, Baroness Elsa was a performance artist, poet and sculptor. She wore cakes as hats, spoons as earrings, black lipstick and postage stamps as makeup. She lived in abject poverty surrounded by her pet dogs and the mice and rats in her apartment, which she fed and encouraged. She was regularly arrested and incarcerated for offences such as petty theft or public nudity. At a time when societal restrictions on female appearance were only starting to soften, she would shave her head, or dye her hair vermilion.

Her work was championed by Ernest Hemingway and Ezra Pound; she was an associate of artists including Man Ray and Marcel Duchamp, and those who met her did not forget her quickly.

Yet the baroness remains invisible in most accounts of the early twentieth-century art world. You see glimpses of her in letters and journals from the time, which portray her as difficult, cold or outright insane, with frequent references to her body odour. Much of what we know about her early life is based on a draft of her memoirs she wrote in a psychiatric asylum in Berlin in 1925, two years before her death.

In the eyes of most of the people she met, the way she lived and the art she produced made no sense at all. She was, perhaps, too far ahead of her time. The baroness is now recognised as the first American Dada artist, but it might be equally true to say she was the first New York punk, sixty years too early. It took until the early twenty-first century for her feminist Dada to gain recognition. This reassessment of her work has raised an intriguing possibility: could Baroness Elsa von Freytag-Loringhoven be responsible for what is often regarded as the most significant work of art in the twentieth century?

Baroness Elsa was born Else Hildegarde Plötz in 1874 in the Prussian town of Swinemünde, now Świnoujście in Poland, on the Baltic Sea. When she was nineteen, following the death of her mother to cancer and a physical attack by her violent father, she left home and went to Berlin, where she found work as a model and chorus girl. A heady period of sexual experimentation followed, which left her hospitalised with syphilis, before she befriended the cross-dressing graphic artist Melchior Lechter and began moving in avant garde art circles.

The distinction between her life and her art, from this point on, became increasingly irrelevant. As her poetry testifies, she did not respect the safe boundaries between the sexual and the intellectual which existed in the European art world. Increasingly androgynous, Elsa embarked on a number of marriages and affairs, often with homosexual or impotent men. She helped one husband fake his own suicide, a saga which brought her first to Canada and then to the United States. A further marriage to Baron Leopold von Freytag-Loringhoven gave her a title, although the Baron was penniless and worked as a busboy. Shortly after their marriage the First World War broke out, and he went back to Europe to fight. He took what money Elsa had with him, and committed suicide shortly afterwards.

Around this time the baroness met, and became somewhat obsessed with, the French-American artist Marcel Duchamp. One of her spontaneous pieces of performance art saw her taking an article about Duchamp's painting *Nude Descending a Staircase* and rubbing it over every inch of her body, connecting the famous image of a nude with her own naked self. She then recited a poem that climaxed with the declaration 'Marcel, Marcel, I love you like Hell, Marcel.'

Duchamp politely declined her sexual advances. He was not a tactile man, and did not like to be touched. But he did recognise the importance and originality of her art. As he once said, '[The Baroness] is not a futurist. She is the future.'

Duchamp is known as the father of conceptual art. He abandoned painting on canvas in 1912 and started a painting on a large sheet of glass, but took ten years to finish it. What he was really looking for were ways to make art outside of traditional painting and sculpture. In 1915 he hit upon an idea he called 'readymades', in which everyday objects such as a bottle rack or a snow shovel could be presented as pieces of art. A bicycle wheel that he attached to a stool in 1913 was retrospectively classed as the first readymade. These were a challenge to the art establishment: was the fact that an artist showcased something they had found sufficient grounds to regard that object as a work of art? Or, perhaps more accurately, was the idea that an artist challenged the art establishment by presenting a found object sufficiently interesting for *that idea* to be considered a work of art? In this scenario, the idea was the art and the object itself was really just a memento that galleries and collectors could show or invest in.

Duchamp's most famous readymade was called *Fountain*. It was a urinal which was turned on its side and submitted to a 1917 exhibition at the Society of Independent Artists, New York, under the name of a fictitious artist called Richard Mutt. The exhibition aimed to display every work of art that was submitted, so by sending them the urinal Duchamp was challenging them to agree that it was a work of art. This they declined to do. What happened to it is unclear, but it was not included in the show and it seems likely that it was thrown away in the trash. Duchamp resigned from the board in protest, and *Fountain*'s rejection overshadowed the rest of the exhibition.

In the 1920s Duchamp stopped producing art and dedicated his life to the game of chess. But the reputation of *Fountain* slowly grew, and Duchamp was rediscovered by a new generation of artists in the 1950s and 1960s. Unfortunately very few of his original works survived, so he began producing reproductions of his most famous pieces. Seventeen copies of *Fountain* were made. They are highly sought after by galleries around the world, even though they need to be displayed in Perspex cases thanks to the number of art students who try to 'engage with the art' by urinating in it. In 2004, a poll of five hundred art experts voted Duchamp's *Fountain* the most influential modern artwork of the twentieth century.

But is it true to say that *Fountain* was Duchamp's work?

On 11 April 1917 Duchamp wrote to his sister Suzanne and said that 'One of my female friends who had adopted the pseudonym Richard Mutt sent me a porcelain urinal as a sculpture; since there was nothing indecent about it, there was no reason to reject it.' As he was already submitting the urinal under an assumed name, there does not seem to be a reason why he would lie to his sister about a 'female friend'. The strongest candidate to be this friend was Baroness Elsa von Freytag-Loringhoven. She was in Philadelphia at the time, and contemporary newspaper reports claimed that 'Richard Mutt' was from Philadelphia.

If *Fountain* was Baroness Elsa's work, then the pseudonym it used proves to be a pun. America had just entered the First World War, and Elsa was angry about both the rise in anti-German sentiment and the paucity of the New York art world's response to the conflict. The urinal was signed 'R. Mutt 1917', and to a German eye 'R. Mutt' suggests *Armut*, meaning poverty or, in the context of the exhibition, intellectual poverty.

Baroness Elsa had been finding objects in the street and declaring them to be works of art since before Duchamp hit upon the idea of 'readymades'. The earliest that we can date with any certainty was *Enduring Ornament*, a rusted metal ring just over four inches across, which she found on her way to her wedding to Baron Leopold on 19 November 1913. She may not have named or intellectualised the concept in the way that Duchamp did in 1915, but she did practise it before he did.

Not only did Elsa declare that found objects were her sculptures, she frequently gave them religious, spiritual or archetypal names. A piece of wood called *Cathedral* (1918) is one example. Another is a cast-iron plumber's trap attached to a wooden box, which she called *God*. *God* was long assumed to be the work of an artist called Morton Livingston Schamberg, although it is now accepted that his role in the sculpture was limited to fixing the plumber's trap to its wooden base. The connection between religion and toilets is a recurring theme in Elsa's life. It dates back to her abusive father mocking her mother's faith by comparing daily prayer to regular bowel movements.

Critics often praise the androgynous nature of *Fountain*, for the act of turning the hard, male object on its side gave it a labial appearance. Duchamp did explore androgyny in the early 1920s, when he used the pseudonym Rrose Sélavy and was photographed in drag by Man Ray. But androgyny is more pronounced in the baroness's art than it is in Duchamp's.

In 1923 or 1924, during a period when Baroness Elsa felt abandoned by her friends and colleagues, she painted a mournful picture called *Forgotten Like This Parapluie Am I By You – Faithless Bernice!* The picture included a leg and foot of someone walking out of the frame, representing all the people who had walked out of her life. It also depicts a urinal, overflowing and spoiling the books on the floor, which had Duchamp's pipe balanced on the lip. The urinal is usually interpreted as a simple reference to Duchamp. But if *Fountain* was Elsa's work, then his pipe resting on its lip becomes more meaningful. The image becomes emblematic of their spoiled relationship.

Fountain is base, crude, confrontational and funny. Those are not typical aspects of Duchamp's work, but they summarise the Baroness and her art perfectly. Perhaps more than anything else, it is this that makes a strong case for *Fountain* being her work, which she sent to Duchamp from Philadelphia to enter in the exhibition, and which he took credit for over thirty years later when both she and the man who photographed the original were dead. To add weight to this claim, Duchamp was said to have bought the urinal himself from J.L. Mott Ironworks on Fifth Avenue, but later research has shown that this company did not make or sell that particular model of urinal.

This is not to suggest that Duchamp deliberately took credit for the female artist's work in a similar manner to how the American painter Walter Keane took credit for his wife Margaret Keane's painting of big-eyed waifs in the 1950s. Psychologists now have a better understanding of how the accuracy of our memories declines over time, which they can model with a diagram known as the Ebbinghaus curve of forgetting. In particular, the excitement that accompanies a good idea emerging in conversation with others can often lead to many people thinking that the idea was theirs. Forty years after Duchamp had submitted *Fountain* to the exhibition, it is entirely possible that he genuinely believed that it had been his idea.

If Duchamp's most famous artwork was not conceived by him, how does this affect his standing as an artist?

The key to his work can be found in a posthumous tribute paid to Duchamp by his friend, the American artist Jasper Johns. Johns spoke of Duchamp's 'persistent attempts to destroy frames of reference'. The frames of reference he was referring to were those of the traditional art establishment, where art was understood to be paintings created by the talent of an artist and then presented to a grateful audience. The reason why Duchamp presented mass-manufactured objects as art was to challenge and ultimately undermine the understanding of what a piece of art was.

Duchamp experimented with chance by dropping pieces of string onto a canvas on the ground, and then gluing them in the position they fell. The aesthetic result was produced by luck, not talent. He did this to undermine the idea that an artist could take credit for their work. He was not prolific, but there was a consistent intention in his work after he abandoned painting. He continually explored the idea that art could not be understood as the product of an artist, for as he said in 1957, 'The creative act is not performed by the artist alone; the spectator brings the work in contact with the external world by deciphering and interpreting its inner qualifications and thus adds his contribution to the creative act.' His 'persistent attempts to destroy frames of reference' were necessary to reveal the spectator's role in the existence of art and to demonstrate that what is observed is in part a product of the observer.

All this leaves us in a strange position. If Duchamp did not conceive or produce *Fountain*, even though he thought he did, and if his goal was to reject the simplistic idea that art is what is produced by artists, then does *Fountain* make him a better, or a worse, artist?

'Persistent attempts to destroy frames of reference' were common in the art of the early twentieth century.

Cubism was developed by the painters Pablo Picasso and Georges Braque in the years after 1907. Painting had begun to move away from lifelike representations of its subjects, but few were prepared for these startling and confusing works. They were strange, angular abstracted images in drab and joyless colours. A common adjective used to describe them was *fractured*, for a cubist painting often looked like an image reflected in the pieces of a broken mirror.

In a cubist painting, the artist's starting point was the recognition that there was no true perspective or framework from which we could objectively view and understand what we were looking at. This was an insight remarkably similar to Einstein's. As a result, the painter did not choose one arbitrary perspective and attempt to recreate it on canvas. Instead, they viewed their subject from as many different angles as possible, and then distilled that into one single image. Here the adjective *fractured* is useful, but misleading. It is not the subject of the painting itself that has become fractured, as is often assumed, but the perspective of the observer. That 'fractured'-looking image should not be thought of as a straight representation of a deeply weird object. It was the painter's attempt to condense their multiple-perspective understanding of a normal subject onto a two-dimensional square of canvas.

Cubist painters weren't radical in their choice of subject. Like the painters who came before them, they were more than happy to paint naked women and still-life groupings of fruit and wine. It was not their subjects that made them radical, but their attempt to create a new and more truthful way of seeing the familiar. As Picasso once said, 'I paint objects as I think them, not as I see them.'

This is the opposite approach to the cubist's contemporaries, the expressionists. The most famous example of expressionism is probably *The Scream* by the Norwegian painter Edvard Munch, in which a man standing on a bridge is depicted as a hellish howl of anguish. Unlike cubism, expressionism sticks with one perspective. Yet it can only justify doing so by recognising how subjective that single vision is. It highlights the artist's emotional reaction to their subject, and makes that an integral part of the work. It understands that the painter's vision is personal and far from objective, but it embraces this limitation rather than attempting to transcend it.

This desire to see in a new way is consistent across other art forms of the early twentieth century. Another example was the development of montage in cinema, a technique primarily developed by the Russian director Sergei Eisenstein during the 1920s. Montage removed the natural links of time and space, which usually connected sequential shots, and instead juxtaposed a number of separate images together in a way that the director found meaningful. In *October: Ten Days that Shook the World*, his 1928 film about the 1917 Russian Revolution, Eisenstein intercuts images of Russian churches first with statues of Christ, and then with religious icons from increasingly distant and ancient cultures, including Buddha, tribal totems, and Hindu and Aztec gods. Contemporary Russian religion was, through this montage, revealed as a contemporary expression of a universal religious spirit. He then immediately began intercutting images of the Russian General Lavr Kornilov with statues of Napoleon, and in doing so forced the audience to see the general as part of a similar ancient historic tradition.

Unlike Braque's or Picasso's cubist canvases, Eisenstein's montages played out in time, so he was able to string together different viewpoints into a sequence rather than merge them into one image. Eisenstein used this clash of perspectives to create many different effects in his filmmaking, from the rhythmic to the symbolic.

The atonal music that Arnold Schoenberg, Alban Berg and Anton Webern composed in Vienna from 1908 onwards was just as startling and strange as cubism. Schoenberg rejected the idea that compositions had to be based on a central tone or a musical key.

In traditional composition, a stream of musical notes complement each other in a way that sounds correct to our ears because the pitch of every note is related to, and determined by, the central tone of the key chosen by the composer. Without that central tone, which all the other notes are based on, we become adrift in what Professor Erik Levi called 'the abyss of no tonal centre'. This is similar to Einstein's removal of the Cartesian x, y and z axes from our understanding of space, on the grounds that they were an arbitrary system we had projected onto the universe rather than a fundamental property of it. Without a tonal centre at the heart of Viennese atonal music, the compositions which followed could be something of a challenge.

Developments in music echoed relativity in other ways. Igor Stravinsky, for example, made great use of polyrhythms in his 1913 masterwork *The Rite of Spring*. A polyrhythm is when two different and unconnected rhythms are clashed together and performed at the same time. The effect can be disorientating, as when different perspectives on an object were clashed together in a cubist painting.

Perhaps more than painting or music, the literature of the early twentieth century has a reputation for being wilfully challenging. What is it about the likes of James Joyce's *Ulysses*, Ezra Pound's *The Cantos* or T.S. Eliot's *The Waste Land* that makes them so unapproachable?

In prose, it is the ongoing story or narrative that acts as the reader's touchstone. It doesn't matter whether the story is told solely from the point of view of one of the characters, or takes the more God-like third-person perspective of an omnipotent narrator. Nor does it matter if the story uses multiple narrators, such as Bram Stoker's *Dracula*, which switches between the voices of many different characters in order to advance its plot. *Dracula*'s varying perspectives are not confusing, in part because they are clearly separated in a way that Eliot's are not, but mainly because all those different voices are telling the same story. This ongoing narrative in prose helps us make sense of everything that happens, in a similar way to the central tone in traditional music or the x, y and z axes of Cartesian space.

Writers like Joyce, Eliot or Pound rejected this singular narrative framework. They frequently shifted narrator, but they did so in a very different way to Bram Stoker. In the second part of *The Waste Land*, the poem's voice abruptly switches to a conversation between women in a British pub, concerning the return of a demobbed soldier husband. There is no introduction to these characters, nor do they seem directly related to the rest of the poem. The effect is jarring and confusing, because there is no central narrative from which this switch of scene makes sense.

The original title of the poem was 'He Do the Police in Different Voices', a reference to a line in Charles Dickens's *Our Mutual Friend* where Betty Higden, talking about her son, remarks that 'You mightn't think it, but Sloppy is a beautiful reader of a newspaper. He do the Police in different voices.' That shift between different voices, clearly, was an important part of what Eliot was trying to do. But

The Waste Land is a better title because that change of viewpoint isn't what the poem is about.

It is about death or, more specifically, the awareness of death in life. The Arthurian reference in *The Waste Land* alludes to an arid spiritual state that is not quite death but in no way life. By getting away from the expected touchstone of a consistent narrative, Eliot was free to look at his subject from a multitude of different angles. He could jump through a succession of different scenes taken from a range of different cultures and time periods, and focus on moments which thematically echoed each other.

The bulk of James Joyce's *Ulysses* concerns a stream-of-consciousness account of a day in the life of Leopold Bloom of Dublin. *Ulysses* is regarded as one of the great twentieth-century novels, but even its staunchest supporters would refrain from describing it as a cracking yarn. Joyce's aim, when he sat down to write, was not to produce a good *story*. As he explained to his friend Frank Budgen, 'I want to give a picture of Dublin so complete that if the city suddenly disappeared from the earth it could be reconstructed out of my book.' He was trying to use the medium of a novel to grasp Dublin from every perspective. Using only a typewriter and reams of paper, Joyce was attempting to do to early twentieth-century Dublin what RockStar North, the Scottish developers of the video game Grand Theft Auto V, did to early twenty-first-century Los Angeles. In Grand Theft Auto V every aspect of the city, including its movies and culture, social media and technology, race relations, stock market, laws and business culture, is recreated and satirised. *Ulysses* is admittedly not often compared to Grand Theft Auto V, but I suspect those familiar with both titles will let the analogy stand.

Modernism is now used as an umbrella term to cover this outpouring of innovation that occurred across almost all forms of human expression in the early twentieth century, most notably in the fields of literature, music, art, film and architecture. Movements such as cubism, surrealism, atonal music or futurism are all considered to be aspects of modernism.

It's not a name that has aged well, if we're honest. Describing work from a century ago as 'modern' is always going to sound a little silly. It suggests that the focus of modernism was on the new – on what was then modern. This is true to a point. Cars, aeroplanes, cinema, telephones, cameras, radios and a host of other marvels were now part of culture, and artists were trying to come to terms with the extent that they were changing everyday life.

Certain forms of modernism, such as futurism, were undeniably a celebration of the new. Futurism was an attempt to visually represent and glorify speed, technology and energy. It was a movement with distinctly Italian roots. Italy, a country which includes men like Enzo Ferrari among its national heroes, produced futurist painters who were besotted with a combination of style and speed.

Modernist architecture was another movement which was in love with the new. In architectural terms this meant new materials, like plate glass and reinforced concrete. The architect Le Corbusier talked of houses as 'machines for living in' where 'form follows function.' There was no place for decoration or ornamentation in that worldview. He wrote about taking a stroll across Paris on an autumn evening in 1924 and being unable to cross the Champs Élysées because of the amount of traffic. This was a new phenomenon. 'I think back to my youth as a student,' he wrote, 'the road belonged to us then; we sang in it.' But Le Corbusier was in no way upset by the changes that he saw. 'Traffic, cars, cars, fast, fast! One is seized with enthusiasm, with joy . . . the joy of power,' he wrote. 'The simple and naive pleasure of being in the midst of power, of strength.' For architects like Le Corbusier, the manic new world was a source of inspiration. He did not say if the novelty of not being able to cross the road eventually wore off.

But while futurist artists and modernist architects were clearly thrilled by the startling culture they found themselves in, modernism was not just a reflection on that brave new world. It was not the case that modernist artists simply painted pictures of cars in the same way that they used to paint pictures of horses. Modernist work could critique modern life as much as celebrate it. It could include an element of primitivism, such as Picasso's use of African masks and imagery, which fetishised a natural, pre-industrial life. There was also early twentieth-century art, such as Gershwin's *Rhapsody in Blue* or the Laurel and Hardy films, which were products of their time but which are not considered to be modernist. Modernism, then, was trying to do something more than just acknowledge the time it was created.

Joyce intended his work to be difficult. We can see this in his reaction to the obscenity trial that resulted from an attempt to publish *Ulysses* in Prohibition-era America. *Ulysses* was originally serialised in a New York magazine called the *Little Review*, alongside the poetry of Baroness Elsa von Freytag-Loringhoven. The baroness's poems may now seem more overtly sexual, but it was Joyce's work which was singled out for obscenity.

A later trial, *United States Vs. One Book Called Ulysses*, in 1933, ultimately decided that the work did have serious intent and that it was not pornographic (for, as Judge John Woolsey pointed out, 'In respect of the recurrent emergence of the theme of sex in the minds of [Joyce's] characters, it must always be remembered that his locale was Celtic and his season Spring.') In order to argue for the serious nature of the work, however, Joyce was called on to explain it, and in particular the way its structure echoed the ancient Greek myth it was named after. Joyce was extremely unhappy with this prospect. As he said, 'If I gave it all up [the explanations] immediately, I'd lose my immortality. I've put in so many enigmas and puzzles that it will keep the professors busy for centuries arguing over what I meant, and that's the only way of ensuring one's immortality.'

Joyce wanted to be studied. As he said in an interview with *Harper's* magazine, 'The demand that I make of my reader is that he should devote his whole life to reading my works.' In this respect there is a touch of tragedy in his dying words: 'Does nobody understand?'

Joyce's work is chewy, in the best sense of the word. There is a rhythm to his language, and a playful disregard for established grammar and vocabulary. His prose somehow highlights the gulf between the words on the page and the subject that they describe. Even when his references go over your head and as far as you can tell nothing seems to be happening, there is a rhythm to his language that keeps you reading. But to do so takes concentration, and prolonged concentration at that. Joyce's work was not written for the attention-deficient twenty-first century. This, however, may be the point.

The need for intense concentration recalls the practices of the mystic George Gurdjieff. Gurdjieff, an imposing and impressively mustachioed Russian who was active in the early to mid-twentieth century, believed that most people lived their lives in a hypnotic form of 'waking sleep'. It was possible to awaken out of this state, Gurdjieff taught, into a state of consciousness commonly experienced by meditators and athletes at their peak, but which is frustratingly difficult to describe to those who have not experienced it. It involves a significant increase in focus, understanding, ability, liberation and joy. It is sometimes said to be as different to normal consciousness as waking is to sleeping. Psychologists increasingly acknowledge a state of mind very much like this, which they call 'flow'.

This state of mind is frustratingly rare and hard to achieve. The key to its activation, Gurdjieff believed, was intense, prolonged concentration. He demanded dedication and commitment from his students, who were set tedious tasks such as cutting a large lawn with a small pair of scissors. His pupils had to mentally force themselves to keep snipping away as the monotonous nature of the task caused their egos to attempt mutiny. This, Gurdjieff thought, would force them into the level of concentration needed to nudge them into a state like flow. Or, in his terminology, it would 'awake their full human potential'. It is tempting to wonder if putting down the lawn scissors and reading *Ulysses* for a few hours would have had a similar effect.

The English author Colin Wilson, whose first book *The Outsider* led to his association with the 1950s literary movement known as the Angry Young Men, had experience of this state. It was, as Gurdjieff predicted, triggered by a period of intense concentration. Wilson was driving a car full of students on New Year's Day in 1979, following a lecture in a remote area of Devon. It had been snowing heavily, and the road was extremely treacherous. 'It was hard to see where the road ended and the ditch began,' he recalled, 'so I was forced to drive with total, obsessive attention.' After about twenty minutes of managing to keep the car on the road he noticed a warm feeling building up in his head. This became a state he called 'peak experience', which continued after the journey had been completed. 'The two hours of concentrated attention had somehow "fixed" my consciousness in a higher state of awareness,' he said. 'There was also an immense feeling of optimism, a conviction that most human problems are due to vagueness, slackness, inattention, and that they are all perfectly easy to overcome with determined effort.' An intense period of effort and concentration had resulted in Wilson seeing the world in an entirely different way.

The difficulty of modernist literature also required the reader to enter a state of intense concentration. It is clear from the modernists' own words that effort on the part of the audience was integral to the work. The difficulty was what created the reward, and that reward justified the effort.

The Romantics, reacting against the age of Enlightenment at the end of the eighteenth century, also believed that we should not be bound by the innate subjectivity of our singular vision. 'May God us keep / From Single vision & Newton's sleep!' wrote William Blake in 1802. But the Romantic's methods of demonstrating this were too vague and whimsical, to modernist thinking at least. The modernists believed that not only were they capable of perceiving a higher perspective, they could make their audience experience it too.

It just wouldn't be easy.

*

We have, as should be apparent, a theme emerging.

That theme, seen repeatedly across the wide sweep of modernist culture, was the idea that a single viewpoint was insufficient to fully express or describe anything. It's an idea that is already familiar to us. The core concept behind Einstein's revolution was that there was no single perspective which can be considered to be correct or true, and that our knowledge of our subject is dependent on the perspective we take.

Was modernism the result of creative people being influenced by Einstein? There are certainly examples where this seems plausible. The history of painting was devoid of melting clocks, for example, before the surrealist painter Salvador Dalí produced *The Persistence of Memory* in 1931. That image appeared after the idea that time could be stretched entered popular consciousness. Dalí has denied that he was influenced by Einstein, and has instead credited a melting camembert as the source of that idea. Yet it remains difficult to dismiss Einstein's impact on the subconscious leap Dalí made between seeing a melting cheese and deciding to paint a melting clock.

Einstein's influence on Dalí is plausible because of the dates involved. By 1931, Einstein was a global celebrity and a sense of his ideas, however caricatured, was widespread. This was not the case when he first published his Special Theory of Relativity in 1905. The world of the physical sciences was then much smaller than comparable sciences such as chemistry. Indeed, the world of science in general was tiny compared to the size it would achieve by the end of the twentieth century. Even among those who had read Einstein's paper, there wasn't an immediate recognition of its significance. Its failure to include gravity, for example, marked it out as more of a curiosity than a clear revolution.

Einstein's importance was finally settled with the publication of his General Theory of Relativity in 1915, but mainstream recognition had to wait until the end of the First World War. The tipping point was in 1919, when Sir Arthur Eddington produced the first experimental proof that Einstein was correct. From that moment on, everyone knew Einstein's name.

The great works of modernism also date to this post-war era, so at first glance the argument for Einstein's influence appears strong. But a long list of modernists, including Picasso, Joyce, Eliot, Braque, Schoenberg, Stravinsky and Kandinsky, had all produced work before the war that showed a clear development of their modernist ideas. There is also work by pre-modernists that pre-dates Einstein. Some early still-life paintings by Gauguin, for example, depict a human face in a corner of the canvas which observes the subject of the painting. You can see this in *Still Life with Profile of Laval* (1886) or *Still Life with Fruit* (1888). A still life is a collection of objects being observed, so if a painter wished to show it as it really was then they should show those items being observed.

Einstein and the modernists appear to have separately made the same leap at the same time. They not only recognised that we are bound by relative perspectives, but they found a higher framework, such as space-time or cubism, in which the subjectivity of a single perspective could be overcome. In 1878 Nietzsche wrote that 'There are no eternal facts, as there are no absolute truths.' Einstein and Picasso were both offering their own solutions to Nietzsche's complaint.

The importance of the observer had been recognised. That such a strange idea should play out across both the creative arts and the physical sciences at the same time is remarkable in itself. That people as radically different as Einstein and Baroness Elsa were working on the same problem indicates that a deeper shift was at work. Something big was happening, and its impact on our culture was widespread.

Even more remarkably, the same idea was emerging from the babbling chaos of international politics.

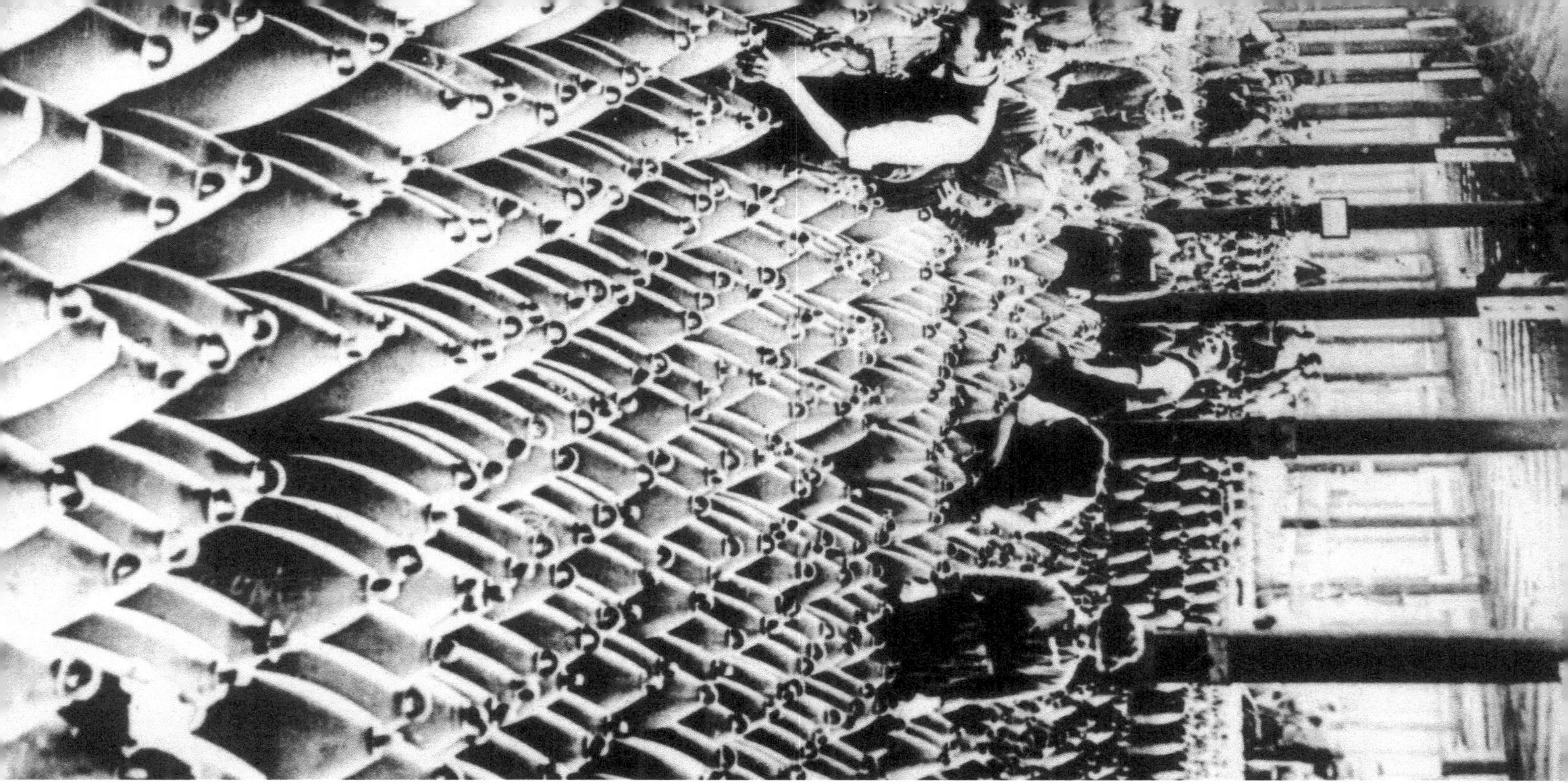

THREE: WAR

The munitions store of a weapons factory, c.1918 (akg)

Hoist that rag

On 17 September 1859 Joshua A. Norton issued a letter to San Francisco newspapers which began, 'At the peremptory request and desire of a large majority of the citizens of these United States, I, Joshua Norton, formerly of Algoa Bay, Cape of Good Hope, and now for the last 9 years and 10 months past of San Francisco, California, declare and proclaim myself Emperor of these United States.'

Norton was an example of the globalisation and migration which became increasingly common in the mid-nineteenth century. He was a British-born, South African-raised businessman who had lost a great fortune in a failed attempt to monopolise the rice market, been declared bankrupt after many years of legal difficulties, and was then living in reduced circumstances in a boarding house. He signed his letter 'NORTON I, Emperor of the United States'. It was duly published and Norton's career as an emperor began.

Following his declaration, Norton began to dress in a blue military uniform with gold epaulettes. He wore a peacock feather in his hat, an imperial sword on his belt, and took to walking with a cane. He added 'Protector of Mexico' to his title but dropped it a decade later, after the realisation that protecting Mexico was a little beyond his powers. The existing photographs of Norton, with his elaborate facial hair and slightly crumpled uniform, show that he somehow managed to look like both a regal emperor and a crazy homeless guy at the same time.

He began issuing proclamations, including one calling for the abolishment of the Republican and Democratic parties, and one which declared that anyone who referred to San Francisco as 'Frisco' would be subject to a twenty-five-dollar fine. Proclamations such as these were as popular with the people of San Francisco then as they no doubt would be now. Although largely penniless, Norton was allowed to eat without payment at the finest San Francisco restaurants, travel for free on municipal transport, and a box was kept for him in a number of theatres. He began issuing his own currency, which was accepted in the bars he used to frequent. When he was once arrested 'for lunacy' the public outcry was such that the police force issued an apology, and took to saluting him when they saw him in the street. He was made a 33rd degree Freemason, and the City of San Francisco provided him with a new uniform when his existing one started to look shabby. When he died in 1880, after more than twenty years as Emperor, thirty thousand people crowded the streets for his funeral. He became immortalised in *The Adventures of Huckleberry Finn,* when Mark Twain based the character of the king on him.

Norton is something of an enigma. It does not appear that he adopted the title as a joke or a scam. He genuinely believed that he was the rightful emperor and thus had a duty to live his life in the manner his status dictated. The public's reaction to Norton points to there being more at play than just the humouring of a crank. In 1969 the American satirical philosopher Greg Hill, the founder and indeed only member of the Joshua Norton Cabal of the Discordian Society, pointed out that 'everybody understands Mickey Mouse. Few understand Hermann Hesse. Hardly anybody understands Einstein. And nobody understands Emperor Norton.'

Norton the First was not, much to his frustration, officially recognised as Emperor of the United States. The closest he came was the 1870 US census, which listed his occupation as 'Emperor', but also noted that he was a lunatic. The United States had an uneasy relationship with the idea of emperors, having been founded in opposition to the British Empire, and up until the mid-twentieth century it stressed a policy of isolationism in its foreign policy. President Wilson successfully ran for re-election in 1916, for example, by stressing his isolationist tendencies with the slogan 'He kept us out of war.' True, the American annexation of the Philippines and other islands following the American-Spanish war of 1898 can only really be described as an imperial act, and few people would claim that the involvement in the separation of Panama from Colombia in 1903, ahead of the building of the Panama Canal, was a textbook example of isolationism. The United States may have been guilty of imperialism in practice, but this did not sit easily with its own national myth. It was not a country that was about to recognise an emperor, no matter how much he looked the part.

There was a self-limiting quirk in the imperial desire of the original east coast American settlers. Their 'manifest destiny' was to spread out across continental North America, regardless of who currently lived on or claimed that land, but only until they reached the Pacific. At that point, once their empire stretched from sea to shining sea, there was to be no more talk of expansion. Their efforts were then focused on quality, not quantity. The aim was to become the biblical 'shining city on a hill'. They wanted to build the best nation that humanity could achieve, not the largest. This made the United States something of an anomaly at the start of the twentieth century.

Outside of the US, at the beginning of the twentieth century, this was still a world of empires and emperors.

The Ottoman Empire stretched across what is now Albania, Macedonia and Turkey, and reached down through Iraq, Syria and Palestine into Africa. Above it lay the Austro-Hungarian Empire. This covered a huge number of what are now modern states, from the Czech Republic in the north down to Bosnia and Herzegovina in the south, and from parts of Italy in the west to parts of Romania and Ukraine in the east. Geographically this was the second-largest European power, beaten only by the vast Russian Empire of Nicholas II to the east. To the north-west lay the German Empire of Wilhelm II and beyond that lay the territory of the elderly Victoria, Queen of the United Kingdom of Great Britain and Ireland and Empress of India. In Imperial China, Empress Dowager Cixi presided over an empire that had held power, in some form, for over two thousand years. The Emperor of Japan, Emperor Meiji, was still alleged to be divine, and had a status not dissimilar to that of ancient Egyptian pharaohs.

Colonialism was a more recent development in the history of empire. The exact definitions of colonialism and imperialism are fiercely debated in academia, but here we're referring to the practice of extending empire into areas of the world that were not geographically adjacent. It first appeared in the fifteenth century, when it became possible to ship well-armed European soldiers far across the world. Spanish and Portuguese rulers laid claim to large parts of South America and Africa. India, which was itself a product of Mughal imperialism, fell under the control of the British during the eighteenth and early nineteenth centuries. By the nineteenth, so many European countries were helping themselves to chunks of Africa that it was considered unusual not to. Colonialism also allowed republics, such as France, to indulge in land grabs that would otherwise be described as imperialistic.

This was how the world had organised itself throughout history. There was scant evidence that other systems were available. As Philip II of Macedon was reputed to have said to his son Alexander the Great in 336 BC, 'My son, ask for thyself another kingdom, for the one which I leave is too small for thee.' Alexander may not have conquered all the known world, but he certainly gave it a good shot. He claimed territory that didn't belong to him through strategy, determination and the possession of a superior army. This was considered admirable and impressive, both in his time and afterwards. He was not viewed, over those long millennia, as a thief, murderer or psychopath, and neither were the emperors who built the Roman, Persian or Egyptian empires. From long before Alexander, emperors were a logical part of human society. It is maybe not surprising that, in an age of unprecedented global migration, Norton's use of the title would have some residual symbolic power in the nineteenth century.

And then, that system ended.

The concept of emperors, so firmly carved into the entirety of world history, collapsed over a few short years. The First World War began on 28 July 1914. By the time the war ended, on 11 November 1918, emperors were discredited beyond redemption. They were the way things always had been, but in a blink they were gone.

Nicholas II of Russia and all of his family were shot dead in a cellar in Yekaterinburg in 1918, after the Bolsheviks seized power. Imperial dynastic China ended with the fall of the Qing dynasty and the establishment of the Chinese republic in 1912, following years of internal strife. The Austro-Hungarian monarchy collapsed at the end of the war in 1918 and the Ottoman Empire was dissolved in 1922. Wilhelm II of Germany avoided being extradited by the Allies for a probable hanging, but he was forced to abdicate and lived the rest of his life in exile. Of all the dynastic Western emperors and empresses who ruled the major world empires at the start of the twentieth century, only the British king remained on his throne when the shooting stopped, and even then the British Empire was wound down and dismantled over the following decades. A similar fate befell the Empire of Japan, which emerged from the First World War intact, but only lasted until the end of the Second.

What caused this sudden change? Before we can begin to disentangle the messy aftermath of the war to end all wars, we're going to take a step back and consider why emperors were so universal, and what changes brought about their downfall. To do this we'll need to go a little further back than you might expect from a book about the twentieth century. But bear with me, for the question of why such a long-lasting system of human organisation should end so abruptly has surprising parallels to the work of the scientists and artists we have already discussed.

Imagine that human societies can be represented by a linear, progressive scale, with those societies increasing in complexity as population grows.

At one end of the scale are small roving bands of hunter-gatherers, which may only number a few dozen. These groupings have no hierarchies or leadership structures other than those that emerge from the normal politics of family life. Property and decision-making are shared, and formal structures are unnecessary.

When the size of the group swells from dozens to hundreds, these bands become tribes. But although decision-making remains egalitarian, there starts to emerge a 'big man' in the tribe who, although not possessing any formal status, tends to be actively involved in conflict resolution and planning. His (or on occasions, her) role is the result of personal aptitude and character. It is a case of the best person for the task stepping up when needed. That person would not receive any special rewards for their actions. They would dress the same as everybody else, perform the same amount of work and live in similar huts or dwellings.

There is considerably more scope for discord when tribes of hundreds grow into chiefdoms which number thousands of people. This was no longer a situation where individuals knew most of the people they saw daily. As the anthropologist Jared Diamond has noted, 'With the rise of chiefdoms around 7,500 years ago, people had to learn, for the first time in history, how to encounter strangers regularly without attempting to kill them.'

In groups of this size there was increased specialisation in work and less equality in wealth. There was now need for a 'chief', a formal role distinct from the rest of the group. The chief would dress differently, live in a more luxurious manner and, on a symbolic level, represent the chiefdom itself. They would make decisions on behalf of the group, and they would frequently be privy to information, such as the ambitions of a nearby tribe, that the rest of the group did not share. The chief ruled with the consent of their group and could usually be replaced at any point, in a similar way to a seventeenth-century pirate captain or a modern-day leader of a gang of bikers.

As populations increased, forms of currency and trade emerged, and the chief's power of decision-making allowed them to become rich. Once power became associated with wealth and privilege, many were willing to take the burden of responsibility from their chief's shoulders. If a leader was to survive, it was necessary to generate popular support for their rule. Cultural and religious structures were a great help here, as were hereditary principles and heavily armed bodyguards. But what really mattered was the principle of protection. The people would support a leader who protected their interests from both internal and external threats. To do this, the chief (who would by now be going by a more grandiose title, such as lord, king or sultan) needed to offer law.

This was the bargain between the ruler and the ruled, which the French called *noblesse oblige*. It was the understanding that with privilege came responsibility. Should a lord provide stability, safety and just and equal law, then the people would pledge him their loyalty in return. This loyalty created legitimacy, which allowed rulers to pursue the wealth, power and prestige they craved. What kings really liked were wars that allowed them to subdue other kings. When kings had other kings under their protection, it allowed them to use an even more impressive title, such as emperor, kaiser or tsar.

Empires were not always popular, especially expensive and unjust ones, but they had some benefits. The single lord ruling over a large territory made all the usual local quarrels and power struggles irrelevant, and the result could be periods of stability and growth. As the revolutionary group the People's Front of Judea ask in the Monty Python movie *Life of Brian*, 'What have the Romans ever done for us? . . . Apart from better sanitation and medicine and education and irrigation and public health and roads and a freshwater system and baths and public order?'

The big problem with empires was that people were not regarded as individuals. They were instead defined by their role in the great imperial hierarchy. They were expected to 'know their place'. If you happened to find yourself in the role of a serf or a peasant rather than a lord or a master, there really was very little you could do about it. This became increasingly problematic following the rise of Enlightenment thought in the late seventeenth century, and the growing acceptance of both rationality and the rights of man.

The flaw in the system became apparent when rulers failed in their duty to provide law, stability and justice. When that occurred in a small chiefdom it took a particularly formidable bodyguard to keep that chief on the throne. Or indeed, to keep that chief's head on top of his body. Yet when that ruler was an emperor with whole armies at his disposal, replacing him was considerably more difficult. A system that worked well for smaller groups became fundamentally flawed at larger scales.

After kingdoms reached the scale where people were no longer able to remove their leader, the incentive to rule justly became less pressing. Rulers became attracted to doctrines such as the divine right of kings. This claimed that a king was not subject to the will of the people, because their right to rule came directly from God. When people believed in the divine right of kings, all notions of egalitarian leadership were over. Leaders were intrinsically superior to their people, at least in their own eyes and in the eyes of those who benefited financially from their power. It is noticeable that leading theologians had very little to say about the divine right of pig-farmers.

This system continued into the twentieth century. It may have been tempered by the growth of representative parliaments, but it was emperors who led the world towards the First World War.

The nineteenth century had been relatively peaceful, at least in Europe. The wars that did occur after the defeat of Napoleon, such as the Franco-Prussian War, the wars of Italian independence or the Crimean War, were brief. Most were over in a matter of months, if not weeks. The only major prolonged conflict was in North America, and that was an internal civil war rather than an expansionist imperial one. When the British Foreign Office announced in August 1914 that Britain was at war with Germany, and huge crowds gathered outside Buckingham Palace to cheer the king in response, there was little reason to think this war would be any different. Although some politicians nursed private fears about a prolonged war, the words of a single British soldier, Joe Armstrong of the Loyal North Lancashire Regiment, summed up the thoughts of many enlisting Europeans: 'Well, I thought the same as everybody else. Everybody said "It'll be over by Christmas and you've got to get out soon, otherwise you won't see anything."'

The jubilant scenes at enlisting stations showed enormous popular enthusiasm for the fight. In Britain, the formation of 'pals battalions', where friends from factories, football teams or other organisations could sign up together and be placed in the same unit, added to the sense that the war would be a bit of an adventure. The schoolboy-like names of these pals brigades, such as the Liverpool Pals, the Grimsby Chums, the Football Battalion or Bristol's Own, make the tragedy of their fate even more acute. The tone of recruitment posters ('Surely you will fight for your King and country? Come along boys, before it's too late') seems horribly disconnected from the horrors that were to come, as does the practice by British women of handing white feathers to men not in uniform to mark them as cowards. Britain had traditionally relied on professional soldiers to fight its wars, so Parliament's plea in August 1914 for a volunteer army of 100,000 men was unparalleled. By the end of September over 750,000 men had enlisted. Those soldiers had no concept of tanks, or aerial warfare or chemical weapons. They did not imagine that war could involve the entire globe. The events to come were unprecedented.

This enthusiasm might now seem bizarre in light of the weakness of the justification for the war. The British were going to war to defend Belgium, which was threatened by Germany's invasion of France and Russia, which was triggered by Russia declaring war on Austria, who were invading Serbia following the shooting of an Austrian by a Serbian in Bosnia. It was a complicated mess, and historians have spent the century since arguing about why it happened. Some have pointed the finger at German imperial expansionism, most notably the German historian Fritz Fischer, yet imperial expansionism was an area in which most of the other combatants were not entirely innocent. The Cambridge historian Christopher Clark argues that the protagonists of this conflict were sleepwalking into the abyss, 'blind to the reality of the horror they were about to bring into the world'. There is no single villain we can blame for what happened. As Clark notes, 'The outbreak of war in 1914 is not an Agatha Christie drama at the end of which we will discover the culprit standing over a corpse in the conservatory with a smoking pistol . . . Viewed in this light, the outbreak of war was a tragedy, not a crime.'

The initial shooting that led to the conflict was itself a farce. The assassin in question was a Yugoslav nationalist named Gavrilo Princip. He had given up in his attempt to kill Archduke Franz Ferdinand of Austria following a failed grenade attack by Princip's colleague, and gone to a café. It is often said that he got himself a sandwich, which would surely have been the most significant sandwich in history, but it seems more likely that he was standing outside the café without any lunch. By sheer coincidence the Archduke's driver made a wrong turn into the same street and stalled the car in front of him. This gave a surprised Princip the opportunity to shoot Ferdinand and his wife Sophie, Duchess of Hohenberg. Over 37 million people died in the fallout from that assassination.

Europe plunged the world into a second global conflict a generation later. The Second World War produced art and literature that were resolute, determined and positive, from songs like 'We'll Meet Again' to movies like *The Dam Busters* or *Saving Private Ryan*. They highlight a clear, unarguable sense of purpose, based around the central understanding that fascism had to be stopped whatever the cost. The First World War, in contrast, produced novels such as *All Quiet on the Western Front* by Erich Maria Remarque or the war poetry of Siegfried Sassoon and Wilfred Owen, all of which examined the war from a point of shocked, uncomprehending horror. The soldiers of the First World War had no comparisons in history to turn to for an explanation of what they had experienced. Remarque fought on the opposing side to Sassoon and Owen, but the questions that these soldiers wrestled with were the same. The experience of the war seems universal, regardless of which side of the trenches a soldier was on and regardless of whether it was recounted by an upper-class poet like Sassoon or a working-class war poet like Ivor Gurney. Much of the most important work did not appear until decades after the conflict, as people were still trying to make sense of their experience of war long after it ended.

This difference in tone is highlighted by two classic war movies, which both tell a broadly similar story of captured officers attempting to escape from a prisoner of war camp. The names of these two films are enough to express their differing character. John Sturges's 1967 film about Second World War Allied prisoners is called *The Great Escape*. Jean Renoir's 1937 story of French First World War prisoners is called *La Grande Illusion*.

With the exception of airmen such as the Red Baron, who won eighty dogfights up in the clouds far removed from life in the mud and trenches, the First World War did not generate popular, romanticised stories. It is instead remembered with static visual symbols – poppies, muddy fields, silhouetted soldiers, trenches, graves – rather than narrative. The closest it came were the spontaneous, unofficial Christmas truces that saw men from both sides leave the trenches, fraternise and play football together. What marked this incident as memorable was that it was not war itself, it was the opposite of war. This ceasefire has become the popular folk memory of the Great War, for who could romanticise the events of Gallipoli, Passchendaele or the Somme? The pointlessness of the conflict can be seen in the stoical humour of the soldiers, who would march to the trenches singing, to the tune of 'Auld Lang Syne', 'We're here because we're here, because we're here, because we're here . . .'

Remarque, Sassoon and other writer-soldiers, and writer-nurses like Vera Brittain, did not don their uniforms to profit themselves. They did so because their king, kaiser or emperor told them to. Most were patriotic and loyal, and enlisted on the back of a wave of enormous popular support for the conflict. As the war dragged on past Christmas 1914 the belief that what they were doing was worthwhile began to falter. By 1917, it was gone. Although early war poems did deal with the expected notions of honour and glory, such as Rupert Brooke's 'The Soldier' ('If I should die, think only this of me: / That there's some corner of a foreign field / That is for ever England'), those poet-soldiers abandoned that approach when the reality of the war became apparent. Assuming that, unlike Brooke who died in 1915 on his way to Gallipoli, they lived long enough.

＊

Why was the reality of the First World War so different to initial expectations? Why was it not over by Christmas, in line with most contemporary assumptions about the nature of conflict? The answer, in part, is technology. It was the first industrialised war.

Before the twentieth century, technology was understood as progress. There had been some protests about the impact of new inventions, most notably by the early nineteenth-century Luddites who destroyed industrial machinery to protest at the effect machines were having on traditional industries. More usually technological advance was believed to be a positive force, providing economic growth and proving mankind's mastery over the natural world. Technology increased what we were capable of. Steam engines allowed us to move great weights, motorcars and bicycles allowed us to travel faster, and telescopes or microscopes allowed us to see what our eyes alone could not. Technology was a tool that provided power or precision, and it did our bidding. But around the beginning of the twentieth century technology started to shrug off our control. Disasters like the sinking of RMS *Titanic* on her maiden voyage, or the uncontrollable fire that consumed the German passenger zeppelin the Hindenburg, showed the downside of progress. Technology could now produce human disasters which were just as terrifying as natural ones. The pseudo-science of eugenics, which aimed to 'improve' the quality of the human race by favouring certain genetic traits, revealed that progress cared little for human emotions like empathy or concern for others.

Professional soldiers in the First World War went to the front trained in the traditional military skills of horse riding and swordplay, but cavalry was soon replaced by tanks, poison gas and machine guns. Career soldiers were soon outnumbered by conscripts and volunteers. Soldiers were not galloping heroically over the fields towards their enemy, but hiding in sodden trenches which did not move for months or even years, along with rats and a terrible lack of food and supplies. And then there was the shelling.

The troops lived under bombardments which could go on for

hours or days or weeks, from the ear-splitting booms of close shelling to the low, bass grumbles of far distant explosions. Each shell arrived unannounced, as if from nowhere. The next shell always had the potential of being a direct hit. Bodies and body parts were lost among the mud and craters, only to resurface again after a later bomb fell. One legacy of shelling was the many Tombs of Unknown Soldiers around the world. These tombs were built after the war, containing anonymous bones which symbolised all lost soldiers. Grieving families paid their respects to remains that could be anyone, such was the ability of this conflict to dehumanise war.

The word 'shellshock' was coined to describe the psychological breakdowns caused by this trauma. This was poorly understood at the time, with some dismissing it as cowardice or 'low moral fibre'. We have since become familiar with the phenomenon, whose symptoms range from near-catatonia to panicked flight, and call it 'combat stress reaction'. Put simply, technology had made warfare psychologically too terrible for soldiers to bear. It took just a few short years for the jubilation of the recruiting stations to become a determination that global conflicts such as this could never be allowed to happen again. This point was hammered home by the name that the conflict soon became known by: the War to End All Wars. When people dared to imagine that such an ingrained constant of history as war would never occur again, then we had clearly entered a new phase in the psyche of mankind.

It was scale that both created and ended the imperial world. Empires were born when population growth caused egalitarian structures to break down. They ended when technology had grown to the point where warfare could no longer be tolerated. The imperial system, it turned out, was not the unarguable, unavoidable system of human organisation that it had been believed to be for most of history. It was a system that only functioned during a certain period of human and technological growth.

If warfare was no longer acceptable in the industrialised world then emperors, tsars and kaisers could no longer be trusted with the power they held. They had stupidly led the world into horror once, and could do so again. The concept of emperors, one of the great constants in human history, was finished. It is impossible to imagine that a modern-day Emperor Norton would receive free food, clothes and travel if they claimed that title after the First World War.

The traditional method of regicide was not hanging or burning, but beheading. If you intended to kill a king it was necessary to chop off his head, as Charles I of England and Louis XVI of France unhappily discovered. This was highly symbolic. It was not just their actual head you were hacking off, but the head of the political hierarchy. An absolute monarch was the omphalos around which the rest of society orientated itself. Squabbles with ministers aside, law and sometimes religion were whatever that emperor decreed them to be. You might not have liked what an emperor did, but you understood that power was theirs. You knew what your role in the hierarchy was, and you orientated yourself accordingly. Without the omphalos of emperors, society was a jumble of different, relative, individual perspectives, all fighting for credibility and political power.

This is what was so remarkable about the changes that occurred in the first decades of the twentieth century. The sudden departure of emperors across huge swathes of the planet was the removal of single, absolute, fixed perspectives. This is a story we've already seen played out in different arenas. Art, physics and geopolitical structures all underwent similar revolutions around the same time, for seemingly unconnected reasons. Politicians were wrestling with the same challenges that faced Einstein, Picasso, Schoenberg and Joyce: how can we proceed, now that we understand there is no ultimate perspective that every other viewpoint is subservient to? How do we reconcile contradictory positions? When our previous ways of thinking are fundamentally flawed, how do we move forward?

These developments would have no doubt pleased an anarchist like Martial Bourdin, who blew himself up by the Greenwich Observatory, had he lived to see them. But at the time when empires were falling, few people thought that the politics of anarchism were a plausible way to organise society. The immediate requirement was stability. Deleting an omphalos leaves you with the chaotic noise of multiple perspectives. Managing this requires a system like democracy.

Limited forms of democracy had been growing for centuries but the vote had been typically limited to elite members of society, such as landowners. Now the long campaigns for universal suffrage, the right for every adult citizen to vote regardless of education, gender or wealth, were about to bear fruit. Universal suffrage appeared in much of Europe, including Norway, Sweden, Austria, Hungary, Poland and the Netherlands, in 1918 or 1919. In the United States the date was 1920, although some ex-Confederate states introduced racial exceptions that were later deemed unconstitutional. Female suffrage often took longer than male, for example in the United Kingdom which gave all adult males the vote in 1918 while women waited until 1928. In countries such as France, Argentina and Japan women had to wait until the end of the Second World War. But for those countries that did not take the communist route out of the failing imperial world, the trend was clear.

Power could not be entrusted to absolute rulers in a world capable of industrialised warfare. The multiple perspective of democracy was safer than the single vision of an emperor. With those emperors gone, political power was redistributed into the hands of individuals.

FOUR: INDIVIDUALISM

Aleister Crowley (third from right) *with companions on an expedition, 1902*
(Picture Post/Hulton Archive/Getty)

Do what thou wilt

In April 1904 the British poet, mountaineer and occultist Aleister Crowley dictated a book which, he believed, was transmitted to him by a non-human intelligence called Aiwass. Aiwass was his Holy Guardian Angel, 'a Being of intelligence and power immensely subtler and greater than aught we can call human'.

Accounts of people who claim to receive information from spirits, angels, strange beings or other non-human sources are common throughout history, and the early twentieth century was no different. In 1913 the Swiss psychiatrist Carl Jung saw visions and heard voices of a being called Philemon. Philemon appeared as an old man with bull's horns and kingfisher's wings, and engaged Jung in deep discussion about the nature of the mind. In 1925 the Irish poet W.B. Yeats and his wife Georgie used automatic writing to contact spirits, who would announce they were ready to communicate by filling Yeats's house with the scent of mint.

When you look at the information received by Crowley, Yeats and Jung it does look remarkably like the work of Crowley, Yeats and Jung. The sources of those communications do not seem to be entirely external, unless non-human entities make contact with mankind in order to do sarcastic impersonations of their psyches. Why such beings would have such a weird sense of humour is a question as yet unanswered.

The book Crowley believed he was transcribing is generally known as *The Book of the Law* because its proper title, *Liber AL vel Legis, sub figura CCXX, as delivered by XCIII=418 to DCLXVI,* is less snappy. It consists of three chapters, each written over the course of an hour, during three days in a hotel room in Cairo. The text is powerful, unsettling and, in places, frightening. The words are short and blunt.

This gives the book an insistent, staccato tone, especially when read out loud. The monosyllabic style is evident in its most famous line: 'Do what thou wilt shall be the whole of the Law.'

Crowley had arrived in the hot, bustling streets of Cairo with his wife Rose in February 1904. The ancient city of mosques and citadels was then undergoing a period of growth and modernisation, and parts of downtown Cairo had been rebuilt in a Parisian style. These wide elegant boulevards with gas lighting stood in stark contrast to the narrow twisting streets and crowded markets of the rest of the city.

This was the peak of the colonial period and Cairo, which was then administered by the British, was regarded as a great prize. Wealthy European travellers were drawn by the romance of the place and the desire to explore the Giza pyramids, the only one of the seven wonders of the ancient world still standing. The purpose of Crowley's visit was to study local religion, play golf and to walk around wearing a jewelled turban, silk robes and a golden coat pretending to be a Persian prince. It was unexpected, therefore, when Rose announced she was channelling strange gnomic information regarding the god Horus, and that Horus was eager to have a word with her husband.

It was all the more surprising because Rose had no interest in Egyptian religion. In order to test her knowledge Crowley took her to the Boulak Museum and challenged her to locate an image of Horus. Rose walked past a number of obvious representations of this deity and headed upstairs. There she immediately pointed to a small display case in the distance and cried, 'There he is!' The pair approached the case and found an otherwise unremarkable wooden stele from the twenty-sixth dynasty. The stele did indeed depict Horus, in the form of Ra Hoor Khuit. It was part of an exhibit that had been numbered 666 by the museum.

This struck Crowley as hugely significant. The number 666 was important for him, and he referred to himself as 'The Beast 666'. So when Rose announced that she had some instructions for Aleister, he followed them to the letter.

Crowley was ordered to arrange their drawing room into a sparse temple and enter it for one hour, starting at noon exactly, on the following three days. There he was to write down exactly what he heard. On each of the three days he sat at a writing table and transcribed the words he heard from an English voice with a neutral accent that sounded like it came from behind him. The voice was of 'deep timbre, musical and expressive, its tones solemn, voluptuous, tender, fierce, or aught else as suited the moods of the message'. This was the voice of Aiwass, an entity who acted as a minister for Horus. On each of the three days he transcribed one chapter of *The Book of the Law*.

As far as Crowley was concerned, *The Book of the Law* marked a new stage in humanity's spiritual evolution. Mankind was entering a period which he called the Aeon of Horus. The previous era, the Aeon of Osiris, was patriarchal in character and echoed the imperial age. You were expected to understand your place in the hierarchy and obey your superiors. The Aeon of Horus, in contrast, was described as being like a child: wild, spontaneous and self-centred. It would be a time when the exercise of individual will was paramount because, as Aiwass dictated, 'There is no law beyond Do what thou wilt.'

Crowley was announcing a new, replacement omphalos. It was one which would come to define the twentieth century: the individual. When the bonds of hierarchy were shattered, you were left with the multiple perspectives of a host of separate individuals. In the philosophy of individualism the self is the focus and is granted precedence over wider society.

Support for individualism had been slowly building for centuries. You can trace its roots back to the Renaissance or the English Civil War. It was boosted by the Enlightenment and can be found in the work of writers such as François Rabelais or the Marquis de Sade. But few people had been willing to take it to its logical conclusion.

Crowley was moving beyond the tradition of Christianity he was raised in. Christianity, like imperialism, was a system of subservience to a higher Lord. This Lord protected and saved his followers, but he also threatened them with judgement and punishment if they did not behave in the manner he dictated. It was a spiritual mirror to the political system: as above, so below. It is no coincidence that Western kings and emperors, from the Holy Roman Empire onwards, went to such lengths to impose that particular religion. Seen in this context, the end of the imperial world was always going to impact the Western world's spiritual models.

Crowley's religion, which he called Thelema, was a product of this new era. Thelema was very different to Christianity in that it did not demand the bending of the knee to anyone. In the words of *The Book of the Law*, 'Every man and every woman is a star.'

In declaring the primacy of the individual, Crowley was also reducing the importance of the social groups that a person belonged to or identified with. Individualism was, by definition, isolating. Focusing on the individual inevitably meant that the focus shifted away from wider social connections, and everything outside the individual became categorised as separate and different. For those who self-identified as being one of the good guys, it became tempting to view this external other as bad. This isolating aspect of individualism was something Crowley understood completely. 'I am alone,' he wrote, 'there is no God where I am.'

The importance of individualism found particularly fertile soil in the United States, which as we have noted was always uncomfortable with the rigid hierarchy of empire. You can see how deeply individualism is ingrained in the American psyche by the unwillingness of American town planners to embrace European-style motoring roundabouts. Roundabouts are faster, cause fewer accidents and save fuel in comparison to traffic-light junctions, but they are viewed as being suspiciously un-American. As Dan Neil, the motoring correspondent at the *Wall Street Journal*, has noted, 'This is a culture predicated on freedom and individualism, where spontaneous cooperation is difficult and regimentation is resisted . . . Behind the wheel, we're less likely to abide by an orderly pattern of merging that, though faster for the group, may require an individual to slow down or, God forbid, yield.'

The central character in F. Scott Fitzgerald's 1925 novel *The Great Gatsby* is James Gatz, who grows up dirt-poor in a shack in North Dakota and changes his name to Jay Gatsby at the age of seventeen in an attempt to leave behind his roots. While serving in the First World War he meets Daisy Buchanan, a debutante from the opposite end of the social scale. Gatsby refuses to accept the social gulf between them and, through an act of individual will and the proceeds from illegal bootlegging, reinvents himself as a wealthy and admired member of the Long Island elite.

Throughout the novel Gatsby stares at a green light across the bay from his home, which marks the Buchanans' estate and the idealised, upper-class society it represents. He fixates on this light, turns it into his own private omphalos, and devotes his life to reaching it. It is this yearning to reach, and to deny the right of any social structure to prevent him from becoming what he wants to be, that is at the heart of his character. Gatsby refuses to 'know his place' or to allow anyone else to determine his goals. His dreams, ultimately, are not sufficient to overcome his background, but his dedication to becoming a 'self-made man' still marks him out as great. His spiritual centre, like Crowley's, rests in his individual will.

The most influential proponent of the argument in favour of taking a fundamentalist approach to personal liberty was probably the Russian-American novelist Ayn Rand. Rand was born Alisa Zinov'yevna Rosenbaum in St Petersburg in 1905. Her childhood was affluent and her father, a successful Jewish businessman, owned both a pharmacy and the building it was in. But when she was twelve years old, Rand's happy childhood was overturned by the October Revolution of 1917. Her father's property was confiscated and her teenage years became a time of uncertainty, desperation and poverty. This left her with a deep hatred of communism, socialism or any type of collectivist ideas. All these, she felt, were excuses to steal from those who earned and deserved their wealth.

After a move to America and a failed attempt to make a living as a screenwriter, she wrote a novella called *Anthem*. *Anthem* describes a dystopian totalitarian future where the word 'I' was banned and had been replaced by 'we'. The hero of her novel, who is initially named Equality 7-2521 but who later calls himself Prometheus, vows to fight this collectivist tyranny. 'I am done with the monster of "We," the word of serfdom, of plunder, of misery, falsehood and shame,' he says. 'And now I see the face of god, and I raise this god over the earth, this god whom men have sought since men came into being, this god who will grant them joy and peace and pride. This god, this one word: "I".' Equality 7-2521 makes this declaration fully aware of the isolating nature of individualism: 'I shall choose friends among men, but neither slaves nor masters. And I shall choose only such as please me, and them I shall love and respect, but neither command nor obey. And we shall join our hands when we wish, or walk alone when we so desire. For in the temple of his spirit, each man is alone.'

Ayn Rand did not believe that concern for the wellbeing of others should limit personal liberty. With her striking short black hair, cold piercing gaze and ever-present cigarettes, she quickly attracted a dedicated following. Her individualist philosophy, which she named Objectivism, promoted what she called 'the virtue of selfishness'. Like Crowley, she viewed her mission as the establishment of a new, post-Christian morality. She made this clear in a 1959 CBS television interview with Mike Wallace, who put it to her that 'You are out to destroy almost every edifice of the contemporary American way of life, our Judeo-Christian religion, our modified government-regulated capitalism, our rule by majority will. Other reviews have said you scorn churches and the concept of God. Are these accurate criticisms?' Rand's response was 'Yes. I am the creator of a new code of morality.'

Crowley, who was by then retired and living in a boarding house in Hastings, Sussex, was a fan. As he wrote in a 1947 letter, a few months before his death, '[Rand's novel] *The Fountainhead* is one of the finest books I have ever read, and my friends in America insist on recognising me in the main character.' In turn, Rand's philosophy would inspire Anton LaVey, the founder of the Church of

Satan. LaVey was the author of *The Satanic Bible*, the most influential text in contemporary Satanism, which has sold over a million copies. LaVey's Satanism was more goat-based than Objectivism, but he readily admitted that his religion was just 'Ayn Rand, with trappings'.

As well as Satanists, Rand also has admirers in the right-wing American Christian and business communities. Ronald Reagan was an admirer. Alan Greenspan, who would spend nineteen years as Chairman of the Federal Reserve, was a member of her inner circle. The Republican Congressman Paul Ryan said in 2005 that 'I grew up reading Ayn Rand and it taught me quite a bit about who I am and what my value systems are, and what my beliefs are. It's inspired me so much that it's required reading in my office for all my interns and my staff. We start with [her longest and last novel] *Atlas Shrugged*.'

This overlap between Rand's admirers and Christian America can be hard to understand, but its roots may lie in the difference between American and European Christianity. During the twentieth century, church attendance declined dramatically across Europe, both in the Protestant north and the Catholic south. Going to church went from being the regular practice of the majority of the population to an unusual, niche interest for a small and ageing minority. European Christianity had always been the spiritual mirror of the restricting, hierarchical imperial system, so its twentieth-century decline at the epicentre of imperialism's collapse isn't surprising.

American Christianity was different. It had evolved in a culture of largely European immigrants who possessed both the proactive spirit which caused them to journey to the other side of the world in search of a better life, and also a dislike of the constricting, controlling power structures of post-industrial revolution Europe. American Christianity had, by necessity, evolved into a faith that was more understanding about the desire for individual freedom. While the idea that a Christian could approve of Ayn Rand appears baffling in Europe, and remains suspicious to the majority of American Christians, there nevertheless exists a section of the

American Christian community which can move from the Bible to *Atlas Shrugged* without a problem. Yet 'the virtue of selfishness' is clearly a different philosophy to 'Love thy neighbour as thyself'.

Towards the end of Crowley's life he boiled down the philosophy of his religion Thelema into a clear, simple one-page document, known as *Liber OZ*. It consists of five short paragraphs, and begins: 'Man has the right to live by his own law – to live in the way that he wills to do: to work as he will: to play as he will: to rest as he will: to die when and how he will.' All this sounds highly appealing. Crowley's philosophy remains attractive as he runs through the next three paragraphs, which detail man's right to eat, drink, dwell, move, think, speak, write, draw, dress and love as he will.

Then we reach the fifth section, which bluntly states that 'Man has the right to kill those who would thwart these rights.' Here you may pause and decide that Thelema isn't quite as attractive as it first appears.

This is an extreme example of the big problem with individualism, that of reconciling the contrasting desires of different people. One individual might wish to express their personal liberty in a way that prevents someone else from doing what they want to do. It's all very well a Thelemite claiming the right to kill, but what if their victim doesn't want to be killed?

Crowley and Rand believed that the solution to a clash of competing liberties was the use of force. When someone was stopping you from doing what you wanted, then the strongest will must prevail. This was also the approach favoured by the Italian politician Benito Mussolini, a key architect of fascism.

Mussolini was very open about his intent to force others to accept his will. 'Everything I have said and done in these last years is relativism by intuition,' he said. 'From the fact that all ideologies are of equal value, that all ideologies are mere fictions, the modern relativist infers that everybody has the right to create for himself his own ideology and to attempt to enforce it with all the energy of which he is capable.'

Three years after being elected to the position of prime minister, Mussolini called a halt to Italy's burgeoning democracy and began his rule as a dictator. He presented his dictatorship as an alternative to liberal democracy and the communism of Vladimir Lenin's Bolshevik party in Russia. He viewed democracy as weak and ineffective, an opinion shared by many after the repercussions of the 1929 Wall Street Crash sent economies around the world into deep depression, most notably Germany's.

Mussolini coined the word 'fascism' to describe his politics. The name came from the word *fasces*, which were a symbol of power in Imperial Rome. *Fasces* were a collection of thin sticks bound together into a rod that was considerably stronger than the sum of its parts. The nation would be stronger if its citizens were bound together in accord with the dictator's will.

Mussolini's fascism partly inspired the rise of Adolf Hitler. When Hitler became Chancellor of Germany in 1933, he attempted to transform the German economy through a totalitarian regime that exercised control over every aspect of the lives of its citizens. Totalitarian states may appear to be the antithesis of individualism, but that very much depends on whether or not you happen to be running them. In the eyes of Crowley or Rand, a leader such as Hitler was admirably exercising his individual will. This is another example of how what is observed is dependent on the position of the observer.

Unfortunately for Hitler, he was not the only psychopath intent on absolute power. Joseph Stalin, who ruthlessly consolidated power in the Soviet Union following the death of Lenin, was also intent on Doing What He Wilt. Stalin and Hitler's politics are considered to be opposing ideologies, but their two totalitarian states could be remarkably similar in practice. As the old Russian joke goes, capitalism was the exploitation of man by man, whereas communism was the reverse.

The most important similarity between the two dictators was their willingness to kill thousands, then tens of thousands, then millions of people. Both executed their own citizens not because of what they had done, but because of who they were. It was explained that some were killed for being Jewish or Slavic and some were killed for being bourgeoisie, but the ultimate reason for their deaths was that Russia and Germany had been rebuilt as monolithic states that obeyed the absolute will of Stalin or Hitler.

When the day came for Stalin and Hitler to attempt to impose their will on each other, the result was as dark as any event in history. The clash of these two dictators was symbolically represented by the Battle of Stalingrad, the largest and bloodiest single battle in the history of war. Russian determination to halt the German advance at Stalingrad ultimately turned the tide of the war on the eastern front and prevented the Nazis from reaching the Crimean oilfields, but about a million and a half people were killed in the process.

The clash between Hitler and Stalin takes the use of force to impose individual will to its logical conclusion, and shows it to be doomed and unconscionable. Thankfully, few people took individualism to such extreme lengths. People are generally practical. Even dedicated libertarians don't argue for their right to drive on whichever side of the road they feel like, for example. Demanding the freedom to drive on both sides of the road would not be worth the trouble, especially for those with any emotional attachment to their lives, their car, or to any friends and family who might be travelling with them. Individualism is not usually the pursuit of total freedom, but a debate about how much personal liberty needs to be surrendered.

Individualism can inspire people to remove themselves from tyranny, rather than inflict tyranny on others. When Rosa Parks decided she was not going to give up her seat, as others dictated that she should, on a racially segregated bus in Alabama in 1955, we see that individualism can be bound by an innate sense of morality, and that an individual act can crystallise a larger communal struggle.

Nevertheless, Rand and her followers promoted complete fundamental selfishness as both rational and moral. The followers of Anton LaVey's Church of Satan recognised that selfishness was morally problematic and evil, but openly admitted that this was what they liked about it. Rand's followers, in contrast, took the moral high ground and argued that it was only through dedicated self-interest that mankind could reach its full potential.

A common analogy used to justify selfishness was the 'invisible hand' which, according to the great Scottish Enlightenment economist Adam Smith, guided stock markets. This was a metaphor for the way in which the accumulation of all the individual, selfish actions in the market produced stability and benefited society. Another analogy was how the blind action of natural selection stabilised the larger ecosystem, especially using the gene-centred view of evolution developed in the 1950s and 1960s by the English biologist W.D. Hamilton and others. This view was popularised by the Oxford evolutionary biologist Richard Dawkins in his 1976 bestseller *The Selfish Gene*. Dawkins's title was another metaphor, for he did not believe that genes were consciously acting in a selfish way. Rather, blind acts of genetic replication inadvertently led to a stable and thriving ecosystem.

In the early 1950s the American mathematician John Nash studied competitive, non-cooperative systems, a branch of mathematics called game theory. It modelled the decisions made by self-interested individuals who operated in an arena devoid of trust. Game theory demonstrated mathematically that the pursuit of self-interest was flawed. The invisible hand may guide and protect, but it could also damage. For example, if a fire broke out in a theatre, self-interest would compel individuals to rush for the exit and push aside people in their way. But if the whole audience did this, more people would be hurt than if everyone left calmly. Another example might be a stock-market panic, or a run on a bank.

The manner in which individuals pursuing their own self-interest harm themselves is known, in economics, as the tragedy of the commons. This analogy dates back to pre-industrial England, when shepherds grazed their flock on shared or 'common' ground. Rationally, a shepherd would benefit most by allowing his sheep to eat as much of the grass on the common land as possible, before it was exhausted by the flocks of other shepherds. But the ultimate result of this logic would harm all shepherds, because the common land would be quickly grazed bare. The best long-term result would be to control the exploitation of the common ground in a sustainable way, as this would offer each shepherd benefits that were smaller in the short term but considerably larger over time. Without such a system in place, rational self-interest dictates that the shepherd must act in a way that ultimately hurts him. A more contemporary analogy would be a corporation that desires to reduce its tax liability and employee wage bill, even though such actions would damage the thriving middle class it needs to purchase its products.

Like Rand's 'enlightened self-interest' and Adam Smith's 'invisible hand', Crowley also had a moral justification for his philosophy. Do What Thou Wilt, he stressed, was very different to Do What Thou Like. This was due to the nature of what he called True Will. True Will was distinct from normal desires, and defined as action that was in harmony with the universe at large. It was action that occurred naturally and which was not undertaken for 'lust of result'. As he saw it, by being 'in harmony with the Movement of Things, thy will be part of, and therefore equal to, the Will of God . . . If every man and every woman did his and her will – the true will – there would be no clashing. "Every man and every woman is a star", and each star moves in an appointed path without interference. There is plenty of room for all; it is only disorder that creates confusion.'

This idea is essentially Daoist. Crowley took the ideas of the sixth-century BC Chinese writer Laozi and flavoured them with a Nietzschean, proto-fascist outlook that chimed with the early twentieth century. It also reflected Crowley's own dark worldview, for there were reasons why the press dubbed him the 'wickedest man in the world'. It was one thing for Crowley to call himself 'The Beast', but it's something else when, as in his case, the person who originally gave him that nickname was his mother.

Crowley's shocking absence of compassion can be seen in his doomed 1905 attempt to scale the treacherous Himalayan peak Kangchenjunga, which is the third-highest mountain in the world. Crowley was a significant figure in the history of climbing, although his occult reputation overshadows his achievements in this field. After a troubled ascent with basic clothing and equipment and a good deal of discord between Crowley and the rest of his expedition, and after reaching a height of 6,500 metres above sea level in terrible conditions, a number of climbers and porters decided to turn back. As they descended the mountain one of the porters slipped and caused an avalanche. Four people were killed, despite the efforts of two survivors to dig them free.

During all this drama Crowley remained in his tent at the camp above. He heard their cries, but chose not to respond. Going to assist his fellow climbers was, apparently, not an action that was in accordance with his True Will. Mountaineers need to be able to rely on their fellow climbers, so the idea that a colleague could ignore calls for help was considered especially unforgivable.

Crowley's opinion of compassion is abundantly clear in *The Book of the Law*. 'We have nothing with the outcast and the unfit: let them die in their misery. For they feel not,' he wrote. 'Compassion is the vice of kings: stamp down the wretched & the weak: this is the law of the strong: this is our law and the joy of the world.' Crowley understood that pure individualism was incompatible with empathy. He was aristocratic in his politics and he would have identified more with the Long Island elite from Fitzgerald's *The Great Gatsby* than with the determined and wilful Gatsby himself. His view, as he so clearly wrote, was that 'The slaves shall serve.'

Those who study Crowley's work argue that there is a great deal of value in his writings, as a scientific system for generating changes in an individual consciousness, provided you can get past the theatrical and proto-fascist trappings. Yet it was Crowley's declaration of the predominance of the individual, rather than his insights into consciousness, that has been his real legacy. This was the reason why the BBC included him in their 2002 list of the 100 Greatest Britons of all time. This was why John Lennon included him on the cover of The Beatles' 1967 album *Sgt. Pepper's Lonely Hearts Club Band*, and why he influenced bands including Led Zeppelin, Black Sabbath and The Rolling Stones.

Crowley wanted to be remembered as the prophet of the great religion of his age. In the UK's 2011 census, 184 people listed their religion as 'Thelemite'. This should be seen in the context of a country where 176,632 people listed their religion as 'Jedi'. The vast majority of the population, Crowley would have been crushed to learn, did not feel that becoming Thelemites was an action that was in accordance with their True Will.

Despite how Crowley and Rand are now viewed, their promotion of fundamentalist individualism had a major impact on the twentieth century. The disappearance of emperors and the omphalos they offered had left us with a mass of competing perspectives. Cubists such as Picasso and physicists like Einstein may have been able to navigate such a world, but most people found it difficult. It was easier to make your individual self the focus of your worldview, especially as this could have short-term advantages. In such circumstances, the problems with individualism could be overlooked. So could questions about what an 'individual' actually was.

Crowley and Rand assumed that an individual was a self-contained, rational agent with free will. The true definition turned out to be considerably messier.

Untitled photomontage by an unknown German artist, c.1910 (adoc-photos/ Corbis)

Under the paving stones, the beach

Something remarkable happened during the May 1913 Parisian premiere of *The Rite of Spring*, a modernist ballet scored by the then relatively unknown Russian composer Igor Stravinsky and performed by Serge Diaghilev's Ballets Russes.

Exactly what happened is still debated. The myth which has grown around the performance paints it as a full-scale riot. The music and performance were so groundbreaking, or so the story goes, that the performance was just too shocking to behold. Stravinsky's work was powerful and atavistic, modern and yet primitive, and it drove the wealthy Parisian cultural elite to violence. *The Rite of Spring* was 'The work of a madman . . . sheer cacophony', according to the composer Puccini.

The stage manager Sergei Grigoriev recalled how, after the curtain rose, 'not many minutes passed before a section of the audience began shouting its indignation; on which the rest retaliated with loud appeals for order. The hubbub soon became deafening; but the dancers went on, and so did the orchestra, though scarcely a note of the music could be heard. The shouting continued even during the change of scene, for which music was provided; and now actual fighting broke out among some of the spectators; yet even this did not deter Monteux [the conductor] from persisting with the performance . . . Diaghilev tried every device he could think of to calm the audience, keeping the lights up in the auditorium as long as possible so that the police, who had been called in, could pick out and eject some of the worst offenders. But no sooner were the lights lowered again for the second scene than pandemonium burst out afresh, and then continued till the ballet came to an end.'

The concert hall was full, and the audience included Marcel

Proust, Pablo Picasso, Gertrude Stein, Maurice Ravel and Claude Debussy. There are many eyewitness accounts of what occurred that evening, but these multiple perspectives bring more confusion than clarity to the account. One claims that about forty people were arrested, while others fail to mention the mass arrival of police that this would entail. The relevant police file is missing from the Paris prefecture archives. Some say vegetables were thrown at the orchestra, yet it seems unlikely that a high-society audience would have brought vegetables to the concert. There are also conflicting reports about whether Stravinsky received an ovation at the end, and whether the outrage was spontaneous or planned by a small group of traditionalist protesters. The Théâtre des Champs-Élysées, where the performance was held, was a modern building with a concrete and steel exterior which many French found too Germanic for their liking. A general unease with the emergence of modernism seems to have played a part in the audience's reaction.

The demographic of the crowd was marked by a split between the modern and the establishment. It included, according to the then twenty-three-year-old Jean Cocteau, 'a fashionable audience, low-cut dresses, tricked out in pearls, egret and ostrich feathers; and side by side with tails and tulle, the sack suits, headbands, showy rags of that race of aesthetes who acclaim, right or wrong, anything that is new because of their hatred of the boxes.' This has led to claims that the riot was a form of class warfare. The French composer Florent Schmitt was reported to have been shouting, 'Shut up, bitches of the *seizième!*' at the *grandes dames* of that prestigious area of Paris.

The riot is a great story, but we've wandered deep into the seductive forest of myth. There is no mention of an actual riot in contemporary reviews and newspaper articles about the premiere. The remaining performances proceeded smoothly, and the piece was soon recognised as a classic – 'the most important piece of music of the twentieth century' in the view of the American composer Leonard Bernstein. It may have had a divided and vocal audience on its first night, but no violence or arrests were recorded in 1913.

An actual riot only tells us about the impact of the performance on one particular day. A mythic riot, on the other hand, shows us that the impact of the music transcends that point in time. Myths don't just crop up anywhere. They need something powerful to form around. There are many badly received premieres, but they do not usually inspire tales of rioting. Perhaps the myth that formed around *The Rite of Spring* tells us there was something rare and powerful in the music itself.

The *Rite* came to Stravinsky in a vision. As he described it in his 1936 autobiography, 'I had a fleeting vision which came to me as a complete surprise, my mind at the moment being full of other things. I saw in [my] imagination a solemn pagan rite: sage elders, seated in a circle, watched a young girl dance herself to death. They were sacrificing her to propitiate the god of spring. Such was the theme of the *Sacre du Printemps*. I must confess that this vision made a deep impression on me.' Stravinsky set about inventing a musical language that could capture the deep, animalistic feelings that his vision conjured. As the German choreographer Sasha Waltz later described that music, 'It conceals some ancient force, it is as if it's filled with the power of the Earth.'

The basis of the music was Russian folk tunes, taken in wild new directions. Stravinsky added dissonance and played at times in more than one key. He included wild clashing polyrhythms with unpredictable stresses which startled the brain and provoked a surge of adrenaline. When Stravinsky first played it to Diaghilev the ballet impresario was moved to ask, 'Will it last a very long time this way?' 'Til the end, my dear' was Stravinsky's response. The ballet company, choreographed by Vaslav Nijinsky, were able to match Stravinsky in capturing this uninhibited frenzy. The dancers, dressed in Russian folk clothes, stomped and flailed as if possessed. Yet for all the focus on recreating the ancient and primal there was no sacrifice of young women in pagan Russia, or indeed in most pagan culture of the last millennium bar the Aztecs and Incas. That addition was the product of the early twentieth century.

The visceral thrill of the music was later captured in *Lost Girls* (1991–2006), a highly literate and highly pornographic graphic novel by the British writer Alan Moore and the American artist Melinda Gebbie. Here the premiere is thematically linked to the assassination of Archduke Ferdinand, and the impact of the music causes not a riot among the audience, but an orgy. The mounting sense of sexual abandon is intercut with the descent into the chaos and berserk rage of the Great War.

At the end of the nineteenth century the imagination of the reading public was captured by Arthur Conan Doyle's great fictional detective Sherlock Holmes. Holmes was a symbol of intellect and rationality, and these were qualities that the prevailing culture held in high esteem. Rationality was promoted as an important virtue that would lead to progress and prosperity. But this focus on the intellect neglected another part of the human psyche. The wild and irrational could not be ignored for ever and, having been repressed for so long, its return would be explosive. For Alan Moore and Melinda Gebbie, this point was marked by the premiere of *The Rite of Spring* and a global descent into total war.

We can also see this moment in W.B. Yeats's 1919 poem 'The Second Coming'. Yeats was a Dublin-born poet and mystic with a great love of Celtic romanticism. 'The Second Coming' was inspired by his belief that the Christian era was coming to an end, and that it would be replaced by something uncertain and violent and unstoppable: 'Things fall apart; the centre cannot hold; / Mere anarchy is loosed upon the world, / The blood-dimmed tide is loosed, and everywhere / The ceremony of innocence is drowned; / The best lack all conviction, while the worst / Are full of passionate intensity,' he wrote. 'And what rough beast, its hour come round at last, / Slouches towards Bethlehem to be born?'

In 1899 the sharp-faced Austrian neurologist and father of psychoanalysis Sigmund Freud published *The Interpretation of Dreams*. It was the book that introduced the wider world to Freud's most important preoccupation: the unconscious mind.

In the same way that Einstein didn't 'discover' relativity but found a way to understand it, Freud did not discover the concept of the unconscious. The idea that there is a part of our mind that we are not consciously aware of, but which still affects our behaviour, was evident in nineteenth-century literature such as the works of Dostoyevsky or *The Strange Case of Dr Jekyll and Mr Hyde* (1886) by Robert Louis Stevenson. Freud's insight was to realise that our sleeping mind could be a key to accessing this hidden part of our selves. 'Dreams,' he wrote, 'are the royal road to the unconscious.'

Before Freud, doctors had been largely clueless about how to treat neurosis. The word 'neurosis' covered a wide range of distressing behaviours, such as hysteria or depression, which were outside of socially acceptable norms but which did not include delusions, hallucinations or other hallmarks of the current understanding of mental illness. Neuroses baffled doctors because there didn't appear to be anything causing them. All they could really do was attempt to treat the symptoms.

This wasn't acceptable to Freud. He believed in causality. He worked on the principle that there must be something causing neurosis, even if the doctor or patient was not aware of what it was. If what we were consciously aware of was only a fraction of what was actually occurring in the mind, he realised, the causes of neurosis must therefore exist in the unconscious part. This was a radical stance. Economists and moralists had long assumed that people were rational beings who knew what they were thinking and who were responsible for their own actions.

Imagine an iceberg. If this iceberg represents Freud's initial model of the mind, then the small part above the surface is the conscious mind. This contains whatever you are thinking about or aware of at that particular moment in time. The area of the iceberg around the water level corresponds to what he called the preconscious mind. This contains thoughts that you may not presently be aware of, but which you could access without trouble should the need arise. Information such as your birthday, email address or your route to work can all be found in the preconscious mind.

Some parts of the preconscious mind are harder to access than others, but this does not mean that those memories and experiences are lost for good. Under certain circumstances they can be recalled, such as a chance meeting with an old friend who reminds you of a highly embarrassing incident which, until then, you had successfully managed to forget. The sense of smell is a particularly evocative way of recalling memories otherwise lost to the unconscious.

Most of our iceberg lies under water, and this great mass represents the unconscious. This is the domain of things that we are not only unaware of, but which we don't have any means to gain awareness of. This, Freud decided, must be the part of the mind that contains the root causes of neurosis. Much of his work was dedicated to developing techniques to somehow bring the contents of this dark place into the light of awareness. As well as dream analysis, he developed the technique of free association, in which patients were encouraged to express whatever crossed their minds in a non-judgemental atmosphere.

Freud developed another model of the mind in 1923, which helped him illustrate how neuroses form. He divided the mind into three separate sections, which he called the *ego*, the *super-ego* and the *id*. The id was like a hedonist, seeking pleasure and desiring new experience. It was driven but addiction-prone and it was, by definition, unconscious. The super-ego, in contrast, was like a Puritan. It occupied the moral high ground, remained steadfastly loyal to the laws and social conventions of the surrounding culture, and attempted to limit or deny the urges of the id.

The task of negotiating between the id and the super-ego fell to what Freud called the ego, which roughly corresponded to the conscious part of the mind. In some ways the name 'ego' is misleading, for the word is commonly associated with an exaggerated and demanding sense of the self. Freud's ego attempted to compromise between the demands of the id and super-ego in a realistic, practical manner, and would provide rationalisations to appease the super-ego when the urges of the id had been indulged. The ego acted like a pendulum, swinging back and forth between the id and the super-ego as events changed.

A rigid, hierarchical culture promotes the arguments of the

super-ego, for this was the part of the mind that strove to please its lord or master. The drives of the id, in such an environment, become taboo, for they are seen as working against a well-ordered society. When the imperial world collapsed in the early twentieth century, the psychological hold that an emperor had over the population weakened. At this point, motivations other than obedience to social norms surface. This was the move from Crowley's patriarchal Age of Osiris into his child-focused Age of Horus, and the shock that so affected the audience of *The Rite of Spring*. The result was that, for many, the ego swung away from the super-ego and found itself facing the long-ignored id.

The surrealists were a group of modernist artists who were heavily influenced by Freud. Their aim was to use and explore the great realm of the unconscious mind that he had revealed to them. The light in their paintings was usually stark, clear, slightly unnerving and highly reminiscent of dreaming. Their art depicted surprising, irrational images displayed with a realistic precision, as if giving validity to the products of the dream world.

Salvador Dalí and Luis Buñuel met in 1922, when they were both art students in Madrid. Dalí was a cubist-inspired painter who was becoming increasingly interested in the irrational. Buñuel practised hypnotism and was fascinated by the workings of the mind, but his great love was cinema. Ever since he was mesmerised by a screening of Fritz Lang's 1921 silent movie *Der müde Tod*, he saw cinema as a powerful medium for exploring surrealist ideas. The very act of entering a dark cinema and losing yourself in silver shadows dancing on a wall put the audience into a state somewhere between waking and dreaming. When Buñuel's mother gave him a gift of some money, the pair had the opportunity to make their first silent film.

In discussing ideas for this first film, Dalí and Buñuel rejected anything based on memories or which was clearly connected to the other images in the film. As Buñuel later explained, writing in the third person, he and Dalí only used scenes which, although they 'moved them profoundly, had no possible explanation. Naturally, they dispensed with the restraints of customary morality and of reason.'

Buñuel told Dalí of a dream in which 'a long tapering cloud sliced the moon in half, like a razor slicing through an eye', and Dalí told Buñuel of his dream about 'a hand crawling with ants'. 'There's the film,' said Buñuel, putting those two scenes together. 'Let's go and make it.' They called the finished film, for no clear reason, *Un Chien Andalou* (An Andalusian Dog).

Earlier attempts at translating surrealism to film, notably by Man Ray and Antonin Artaud, had not been successful. Dalí and Buñuel assumed their effort would get a similar reaction. When the film was first screened in Paris, Buñuel, who was playing phonogram records by the side of the screen to provide a soundtrack for his silent movie, kept his pockets full of stones in case he needed to throw things at hecklers. This proved not to be necessary. The film was a triumph and André Breton, the author of the Surrealist Manifesto, formally accepted the Spanish filmmakers into the ranks of the surrealist movement. The film is still screened regularly by film societies to this day, not least because it is short, and surprisingly funny. The opening scene, where a shot of a woman calmly having her eye sliced by a razor is intercut with a thin cloud crossing the moon, is still one of the most striking and powerful images in cinema.

The reaction earned Dalí and Buñuel the opportunity to make a second, longer and more substantial movie, which was financed by a wealthy patron. This was *L'Âge d'or* (1930). A falling-out between the pair meant that Dalí's involvement did not extend beyond the script. Buñuel was not the only significant figure in Dalí's life to break with him at this point. His father threw him out of the house following his exhibition of a drawing entitled *Sometimes, I spit for fun on my mother's portrait*.

Buñuel finished the film himself. As the reviewer in *Le Figaro* put it, 'A film called *L'Âge d'or*, whose non-existent artistic quality is an insult to any kind of technical standard, combines, as a public spectacle, the most obscene, disgusting and tasteless incidents. Country,

family and religion are dragged through the mud.' At one screening the audience threw purple ink at the screen before heading to
a nearby art gallery in order to vandalise surrealist paintings, and
the patrons who financed the film were threatened with excom-
munication by the Vatican. The scandal resulted in the film being
withdrawn. It was not seen again in public for nearly fifty years.

Officially, the reason for the offence was the end sequence. This
referenced the Marquis de Sade's *120 Days of Sodom*, a story about
four wealthy libertines who lock themselves away for a winter in
an inaccessible castle with dozens of young victims whom they
proceed to rape and murder in order to experience the ultimate
in sexual gratification. The book was an exploration of the darker
extremes of the 'Do what thou wilt' philosophy. It was written in
1785 but could not be published until 1905. It was, in Sade's own
estimation, 'the most impure tale that has ever been told since our
world began'. *L'Âge d'or* did not depict any of this depravity, with the
exception of one off-screen murder, but instead showed the four
exhausted libertines exiting the castle at the end of the story. The
problem was that one of the four was depicted as Jesus.

This blasphemy was officially the reason for the great scandal.
But what was actually shocking about the film was its depiction of
female desire. This ran throughout the main section of the film,
and culminated in the unnamed female character, unsatisfied and
frustrated after the casually violent male character leaves, lustfully
licking and sucking the toes of a marble statue.

Not even early cinema pornographers went that far. They may
have stripped women naked, but those women were still portrayed
as coy and playful. Even depictions of predatory women steered
away from such an uninhibited expression of lust. Banning the film
on the grounds of the Jesus scene was like jailing Al Capone for tax
evasion. It was the easiest way to get the job done, but it was clearly
an excuse.

L'Âge d'or was, in the words of the American novelist Henry Miller,
'a divine orgy'. Miller had arrived in Paris in 1930, as the film was
released, and immediately began to write similarly frank depictions

of sexuality. Like James Joyce's *Ulysses*, Miller's first published book, *Tropic of Cancer* (1934), was banned from being imported into the United States by the US Customs Service on the grounds of obscenity. Prosecutors won that battle, but they were losing the war. Culture was becoming increasingly open about sexuality and there was no hope of prosecuting everybody.

In 1932 the American actress Tallulah Bankhead shocked many when she said, 'I'm serious about love . . . I haven't had an affair for six months. Six months! . . . The matter with me is, I WANT A MAN! . . . Six months is a long, long while. I WANT A MAN!' Those offended by female sexuality had to come to terms with both the fact that she made this statement, which would have been unthinkable a generation earlier, and that *Motion Picture* magazine saw it as fit to print.

The 1920s had been the jazz age, a golden era for the wealthy that in retrospect sat in stark contrast to the world war that preceded it and the grinding global depression that followed. Its archetypal image was that of a flapper, a woman with pearls, a straight dress and a short bob, dancing joyfully and freely, kicking her legs and enjoying herself unashamedly. The reason such a simple image came to define that era was because of how unprecedented it was. Such public behaviour by a high-status female would have been unacceptable at any earlier point in the Christian era.

Flappers were not just accepted, they were celebrated. The black Missouri-born dancer Josephine Baker might have been ignored in her homeland due to her colour, but that was not the case in Europe. Her performances in Paris, dressed in nothing more than a skirt made from feathers or bananas, were wild and blatantly sexual and also incredibly funny. They made her one of the most celebrated stars of her day. Baker loved animals and surrounded herself with exotic creatures, including a snake, a chimpanzee and a pet cheetah named Chiquita. She was showered with gifts from wealthy admirers, and claimed to have received approximately fifteen hundred marriage proposals. Following her death in 1975 she became the first American woman to receive full French military honours at her funeral, thanks to her work with the Resistance during the Second World War.

Jazz music and dances such as the Charleston, the Black Bottom or the Turkey Trot were seen as modern and liberating. Dresses became simpler, and lighter. Skirts became shorter, reaching the previously unimaginable heights of the knee. The amount of fabric in the average dress fell from almost twenty yards before the Great War, to seven. The fashionable body shape was flat-chested and thin, a marked contrast to previous ideals of female beauty. In the nineteenth century lipstick had been associated with prostitutes or, even worse, actresses. By the 1920s it was acceptable for all, and cupid-bow lips were all the rage. In the words of the American journalist Dorothy Dunbar Bromley, women were 'moved by an inescapable inner compulsion to be individuals in their own right'.

The credit for the power of the *L'Âge d'or* toe-sucking scene must go to Buñuel. Dalí was not comfortable with female sexuality. His personal sexuality was focused more on voyeurism and masturbation. He was devoted to his wife Gala, but preferred her to sleep with other men. 'Men who fuck easily, and can give themselves without difficulty, have only a very diminished creative potency,' he said. 'Look at Leonardo da Vinci, Hitler, Napoleon, they all left their mark on their times, and they were more or less impotent.' Dalí was reportedly a virgin on his wedding night, due to his fear of the vagina, and frequently linked seafood with female genitalia or sexuality in his work. His famous 1936 sculpture *Lobster Telephone*, which was a telephone with a plastic lobster attached to the handle, was also known by the alternative name of *The Aphrodisiac Telephone*.

Thanks to his cartoon moustache, his stream-of-consciousness declarations of his own genius and his love of luxury and power, it is tempting to see Dalí's public persona as some form of calculated performance art. But from descriptions of him by those in his inner circle, there does not appear to have been a private Dalí which differed from the public one. 'Every morning upon awakening, I experience a supreme pleasure: that of being Salvador Dalí, and I ask myself, wonderstruck, what prodigious thing will he do today, this Salvador Dalí, he once said. Very few people would allow themselves to say such a sentence out loud.

Dalí did not have the self-consciousness filter that most people employ to present a more socially acceptable image of themselves. To use Freud's model, he lacked the super-ego to stop the id pouring out of him. 'I am surrealism,' he once said, as if his ego was unimportant relative to the work that came through him. Freud was certainly impressed. 'I have been inclined to regard the Surrealists as complete fools, but that young Spaniard with his candid, fanatical eyes and his undeniable technical mastery has changed my estimate,' he wrote in 1939. Others were less impressed. As Henry Miller put it, 'Dalí is the biggest prick of the twentieth century.'

Freud's model of the id, ego and super-ego was originally only intended to describe individuals. But there is a tradition of using Freudian ideas to help illuminate larger changes in society, for example in works like Wilhelm Reich's *The Mass Psychology of Fascism* (1933). Freud's psychological models can be used alongside a sociological concept known as 'mass society', which describes how populations of isolated individuals can be manipulated by a small elite. The idea of the 'mass media' is related to that of the mass society. Methods of controlling or guiding mass society were of great interest to political leaders.

An example of the subconscious manipulation of mass society was the twisting of people's reaction to different ethnicities in the 1930s. When political leaders promoted hatred of others, it created one of those rare instances that appealed to both the id and the super-ego at the same time. It was possible to unleash the barbaric, destructive energies of the id while at the same time reassuring the super-ego that you were loyally obeying your masters. With the id and the super-ego in rare agreement, the ego could find it hard to resist the darkness that descended on society.

When the wild energies of the id were manipulated in a precise way, leaders could command their troops to organise genocides. The word 'genocide' was coined in 1944 to describe a deliberate attempt to exterminate an entire race. It hadn't existed before. There had been no need for it before the twentieth century. Exact numbers are difficult to pin down, but most estimates say that Stalin was responsible for more deaths than Hitler, and that Mao Zedong was responsible for more deaths than Hitler and Stalin combined. Men such as Pol Pot, Saddam Hussein and Kim Il Sung all played a part in ensuring that the twentieth century would forever be remembered as a century of genocide.

The off-hand manner in which genocide was shrugged off is chilling. 'Who still talks nowadays of the extermination of the Armenians?' Hitler told Wehrmacht commanders in a speech delivered a week before the invasion of Poland. Hitler was aware of how the global community had either accepted the Armenian genocide, in which the Ottoman government killed up to a million and a half Armenians between 1915 and 1923, or had turned a blind eye to it. As Stalin was reported to have remarked to Churchill, 'When one man dies it is a tragedy, when thousands die it's statistics.'

Modern technology made all this possible. Hitler kept a portrait of the American car manufacturer Henry Ford on the wall of his office in Munich. Ford was a notorious anti-Semite who had developed assembly-line techniques of mass production based on Chicago slaughterhouses. The application of a modern, industrialised approach to killing was one factor which differentiated modern genocides from the colonisation of the Americas and other such slaughters of the past. But the availability of techniques to industrialise mass killing does not in itself explain why such events occurred.

In 1996 the president of Genocide Watch, Gregory Stanton, identified eight stages that occur in a typical genocide: Classification, Symbolisation, Dehumanisation, Organisation, Polarisation, Preparation, Extermination and Denial. The first of these, Classification, was defined as the division of people into 'us and them'. This was something that the particular character of the twentieth century was remarkably suited towards. It was a side effect of both nationalism and individualism. Focusing on the self creates a separation of an individual from the 'other' in much the same way as identifying with a flag does.

Genocides arose during the perfect storm of technology, nationalism, individualism and the political rise of psychopaths. They revealed that humans were not the rational actors they prided themselves on being, dutifully building a better world year after year. Rationality was the product of the conscious mind, but that mind rested on the irrational foundations of the unconscious. The individual was more complicated than originally assumed. If there was some form of certainty to be found in the post-omphalos world, it wouldn't be found in the immaterial world of the mind.

The next question, then, is whether such certainty could be found in the physical world?

SIX: UNCERTAINTY

Erwin Schrödinger, c.1950 (SSPL/Getty)

The cat is both alive and dead

O n the last day of the nineteenth century, six hours before mid-night, the British academic Bertrand Russell wrote to a friend and told her that 'I invented a new subject, which turned out to be all mathematics for the first time treated in its essence.' He would later regard this claim as embarrassingly 'boastful'.

Russell was a thin, bird-like aristocrat who compensated for the frailness of his body through the strength of his mind. He became something of a national treasure during his long life, due to his frequent television broadcasts and his clearly argued pacifism. His academic reputation came from his attempt to fuse logic and mathematics. Over the course of groundbreaking books such as *The Principles of Mathematics* (1903) and *Principia Mathematica* (1910, co-written with Alfred North Whitehead), Russell dedicated himself and his considerable brain to becoming the first person to prove that $1 + 1 = 2$.

Russell had a solitary childhood. He was brought up in a large, lonely house by his stern Presbyterian grandmother following the death of his parents, and he lacked children of his own age to play with. At the age of eleven, after his older brother introduced him to Euclidian geometry, he developed a deep love for mathematics. In the absence of other children he became absorbed in playing with numbers.

Yet something about the subject troubled him. Many of the rules of mathematics rested on assumptions which seemed reasonable, but which had to be accepted on faith. These assumptions, called axioms, included laws such as 'For any two points of space there exists a straight line that connects those two points' or 'For any natural number x, $x + 1$ is also a natural number.' If those axioms were accepted, then the rest of mathematics followed logically. Most mathematicians were happy with this situation but, to the mind of a gifted, lonely boy like Russell, it was clear that something was amiss. He was like the child in *The Emperor's New Clothes*, wondering why everybody was ignoring the obvious. Mathematics, surely, needed stronger foundations than common-sense truisms. After leaving home and entering academia, he embarked on a great project to establish those foundations through the strict use of logic. If anything could be a system of absolute clarity and certainty, then surely that would be logically backed mathematics.

In the real world, it is not too problematic to say that one apple plus another apple equals two apples. Nor is there much argument that if you had five scotch eggs, and ate two of them, then you would have three scotch eggs. You can test these statements for yourself the next time you are in a supermarket. Mathematics, however, abstracts quantities away from real-world things into a symbolic, logical language. Instead of talking about two apples, it talks about something called '2'. Instead of three scotch eggs, it has the number '3'. It is not possible to find an actual '2' or a '3' out in the real world, even in a well-stocked supermarket. You'll find a squiggle of ink on a price tag that symbolically represents those numbers, but numbers themselves are immaterial concepts and hence have no physical existence. The statement $1 + 1 = 2$, then, says that one immaterial concept together with another immaterial concept is the same as a different immaterial concept. When you remember that immaterial concepts are essentially things that we've made up, the statement $1 + 1 = 2$ can be accused of being arbitrary. Russell's project was to use logic to prove beyond argument that $1 + 1 = 2$ was not an arbitrary assertion, but a fundamental truth.

He nearly succeeded.

His approach was to establish clear definitions of mathematical terms using what logicians then called *classes*, but which are now better known as *sets*. A set is a collection of things. Imagine that Russell wanted a logical definition of a number, such as 5, and that he also had a vehicle with a near-infinite storage capacity, such as

Doctor Who's TARDIS. He could then busy himself travelling around the world in his TARDIS looking for examples of five objects, such as five cows, five pencils or five red books. Every time he found such an example, he would store it in his TARDIS and continue with his quest. If he successfully found every example of a five in the real world then he would finally be in a position to define the immaterial concept of '5'. He could say '5' is the symbol that represents the set of all the stuff he had collected in his magic blue box.

Producing a similar definition for the number 0 was more complicated. He could hardly travel the world filling his TARDIS with every example of no apples or no pencils. Instead, he defined the number 0 as the set of things that were not identical to themselves. Russell would then fill his TARDIS with every example of a thing that was not the same as itself and, as there aren't any such things in the world, he would eventually return from this fruitless quest with an empty TARDIS. Under the rules of logic there is nothing that is not identical to itself, so this was a valid representation of 'nothing'. In mathematical terms, he defined the number 0 as the set of all *null sets*.

If Russell could use similar, set-based thinking to produce a clear definition of 'the number 1' and the process 'plus 1', then his goal of being able to prove beyond doubt that $1 + 1 = 2$ would finally be achievable. But there was a problem.

The problem, now known as Russell's paradox, involved the set of all sets that did not contain themselves. Did that set contain itself? According to the rules of logic, if it did then it didn't, but if it didn't, then it did. It was a similar situation to a famous Greek contradiction, in which Epimenides the Cretan said that all Cretans were liars.

At first glance, this may not appear to be an important paradox. But that wasn't the point. The problem was that a paradox existed, and the goal of rebuilding mathematics on the bedrock of logic was that it would not contain any paradoxes at all.

Russell went back to first principles, and a number of new definitions, arguments and fudges were proposed to avoid this problem.

Yet every time he built up his tower of mathematical logic, another problem revealed itself. It felt like paradoxes were unavoidable aspects of whatever self-contained system mathematicians produced. Unfortunately, that turned out to be the case.

In 1931 the Austrian mathematician Kurt Gödel published what is now known as Gödel's Incompleteness Theorem. This proved that any mathematical system based on axioms, complex enough to be of any use, would be either incomplete or not provable on its own terms. He did this by coming up with a formula which logically and consistently declared itself unprovable, within a given system. If the system was complete and consistent then that formula would immediately become a paradox, and any complete and consistent system could not contain any paradoxes. Gödel's theorem was extremely elegant and utterly infuriating. You can imagine how mathematicians must have wanted to punch him.

This did not mean that mathematics had to be abandoned, but it did mean that mathematical systems always had to make an appeal to something outside of themselves. Einstein had avoided the contradictions in the physical world by going beyond normal three-dimensional space and calling on the higher perspective of space-time. In a similar way, mathematicians would now similarly have to appeal to a higher, external system.

If any branch of thought was going to provide an omphalos which could act as an unarguable anchor for certainty, then common sense said that it would have been mathematics. That idea lasted no longer than the early 1930s. Common sense and certainty were not faring well in the twentieth century.

For people with a psychological need for certainty, the twentieth century was about to become a nightmare.

The central monster in that nightmare was a branch of physics known as quantum mechanics. This developed from seemingly innocuous research into light and heat radiation by scientists at the turn of the century, most notably Einstein and the German physicist Max Planck. This spiralled off into an intellectual rabbit hole so strange and inexplicable that Einstein himself feared it marked the end of physics as a science. 'It was as if the ground had been pulled out from under one,' he said, 'with no firm foundation to be seen anywhere, upon which one could have built.'

Einstein was not alone in being troubled by the implications of this new science. The problem was, in the words commonly attributed to the Danish physicist Niels Bohr, 'Everything we call real is made of things that cannot be regarded as real.' As Richard Feynman, arguably the greatest postwar physicist, later admitted, 'I think I can safely say that nobody understands quantum mechanics.' The Austrian physicist Erwin Schrödinger probably summed up the situation best when he said, 'I do not like [quantum mechanics], and I am sorry I ever had anything to do with it.' But people not liking quantum physics does not change the fact that it works. The computer technology we use every day is testimony to how reliable and useful it is, as the designs of computer chips rely on quantum mechanics.

Quantum mechanics was entirely unexpected. Scientists had been happily probing the nature of matter on smaller and smaller scales, oblivious to what horrors awaited. Their work had proceeded smoothly down to the level of atoms. If you had a pure lump of one of the ninety-two naturally occurring elements, such as gold, and you cut that lump into two pieces and discarded one, what remained would still be a lump of gold. If you continually repeated the process you would still find yourself with a piece of gold, albeit an increasingly small one. Eventually you would be left with a piece that was only a single atom in size, but that atom would still be gold.

It is at this point that the process breaks down. If you were to split that atom into two halves, neither of those halves would be gold. You would have a pile of the bits that once made a gold atom, the discrete *quanta* which give the science its name, but you would not have any gold. It would be like smashing a piñata and ending up with a pile of sweets and broken papier mâché, but no piñata.

At first, things looked neat enough. An atom consisted of a centre or nucleus, which was made up of things we called protons and neutrons. These were orbited by a number of electrons, which were much smaller and lighter than the protons and neutrons. In time, it became clear that some of these bits could be broken down even further. A proton, for example, turned out to be made up of a number of smaller things called quarks. An atom was constructed from a family of a few dozen different building blocks, which soon gained exotic names like leptons, bosons or neutrinos. All these were given the generic name of subatomic particles.

The problem here is the word 'particle'. It seemed like a reasonable word at first. A particle was a tiny object, a discrete thing with mass and volume. Scientists liked to imagine particles as being like snooker balls, only smaller. They were actual things that you could, in theory, put in a cupboard or throw across the room. Physicists measured an aspect of these particles which they called 'spin', which they said could be either clockwise or anticlockwise, as if the tiny snooker ball was rapidly rotating. The classic illustration of the atom was a cluster of snooker balls in the centre, with a few more circling in clearly marked orbits. The study of the subatomic world, it was assumed, was like studying how snooker balls collide and behave, except smaller. Or at least, that's how people instinctively assumed it should be.

But it wasn't.

As research progressed, scientists found themselves in the strange position of knowing a lot about how subatomic particles behaved, but knowing nothing about what they actually were. One suggestion, which has been studied in great detail from the mid-1980s onwards, was that subatomic particles are all we can see of multi-dimensional vibrating strings. We are still unable to say whether or not this idea is actually true. The only thing we know with any certainty is that we don't know what these things are.

We know that these building blocks of atoms are not tiny snooker balls because they also behave like waves. Waves, such as sound waves or waves on the sea, are not discrete lumps of stuff but repeating wobbles in a medium, such as air or water. Experiments intended to show how subatomic particles behaved like waves conclusively proved that, yes, they did indeed behave like waves. But experiments that were intended to show that these things were discrete particles also conclusively showed that they behaved like individual particles. How was it possible that the same thing could behave like tiny snooker balls and also behave like waves? It was like finding an object that was simultaneously both a brick and a song.

Studying objects which were two contradictory things at the same time was something of a challenge. It was like Zen Buddhism with extra maths. It emerged that these subatomic things could be in more than one place at once, that they could 'spin' in different directions at the same time, move instantaneously from one place to another without passing through the distance in between, and in some way communicate instantaneously over great distances in contradiction of all known laws. All this was bad enough without having to assign such behaviour to objects viewed as both waves and particles at the same time. Yet incredibly, following some serious intellectual arguments in the first half of the twentieth century and some really expensive experiments in the second, much of the behaviour of subatomic thingamabobs is now predictable.

One result of the simultaneous acceptance of both the 'wave' and 'particle' models was that these objects were considered to be extremely strange. This could not be more wrong. Their behaviour is the most commonplace and unremarkable thing in the universe. It is occurring, constantly and routinely, everywhere around you, and so is surely the opposite of 'strange'. The reason why we think subatomic particles are strange is because they are so different to how things appear at a human-scale perspective. It is once again down to the observer as much as the observed. It is our problem, not the universe's.

The danger with using wave and particle models is that it is very easy to forget they are metaphors. To grasp the nature of the subatomic world we have to accept that the things in it are neither waves nor particles, as those words are commonly understood. We need a different metaphor, one that recognises that the world at that scale runs on different or special laws.

The late science fiction author Douglas Adams once noted that 'Nothing travels faster than the speed of light, with the possible exception of bad news, which obeys its own special laws.' In tribute to Douglas Adams we will forget waves and particles and adopt news as our metaphor for describing the subatomic world, because there is no danger that we will take this literally.

Let us imagine a single unit of news, such as Vladimir Putin, the President of Russia, being photographed fighting a kangaroo. Such an event is unpredictable, in that it is not possible to say in advance when it is going to occur. All we can say, and both supporters and detractors of Putin will agree on this point, is that at some point the President is going to punch a kangaroo. That's just the sort of person he is.

This news event is analogous to a subatomic event, such as nuclear decay. Nuclear decay is the process in which an atom of uranium or other unstable element spits out radiation. We can calculate averages and predict how much radiation a given chunk of uranium will emit over a certain time, but we are unable to say when the process will occur in a particular atom. Like Vladimir Putin fighting a kangaroo, it might happen right now, it might happen in half an hour, or it might happen in twenty years. There's no way of knowing until it occurs.

This uncertainty was in itself a shocking discovery. Our universe, it was believed, ran under a strict process of cause and effect. It was no good having strict physical laws about what caused things to happen if nature was just going to do stuff whenever it felt like it. The realisation that an atom would decay for no reason at all, at a time of its own choosing, was profoundly unsettling. Einstein himself refused to accept it, famously remarking that 'God doesn't play dice with the world.' Einstein believed that there must be a fundamental reason hidden somewhere deep within the atom to explain why it decays at certain moments and not others. No such mechanism has been found, and the scientific community is now more in agreement with the English theoretical physicist Stephen Hawking, who has said that 'God does play dice with the universe. All the evidence points to him being an inveterate gambler, who throws the dice on every possible occasion.'

But back to our news event.

Now that Putin has had a fight with a kangaroo, what will be the result? We can be certain that this single event will result in a wide range of media coverage. We could calculate this media coverage to a high degree of accuracy, assuming we knew factors such as the political affiliations of editors, bloggers, newspaper or television channel owners and the demographic target of the news outlets. Some coverage would be funny, some disgusted, some sensational and some irate. You could no doubt predict in advance what approach your favoured news source would take.

What, though, would be occurring in the period between the fight and the publication of the resulting news reports? There would be a lot of mental analysis about what had happened, and a lot of different potential interpretations would be considered. Some people might think that the fight made Putin a strong leader, and some would think that it made him a horrible person. Others might wonder if the fight had been faked for the cameras, or was representative of the current state of animal rights in Russia, or whether Putin was having a breakdown. Many would assume that the President was drunk. There would be people who thought that the whole incident was cynical media manipulation, designed to distract attention from an unrelated political scandal. There would probably be some people who entertain an elaborate conspiracy theory in which the kangaroo had been trained to throw the fight, or who would believe that they are oppressed by political correctness and should also be able to punch kangaroos whenever they feel like it. All these thoughts and interpretations would immediately be recorded on social media, alongside jokes, photoshopped images and fake accounts claiming to be Putin's kangaroo.

These thoughts, and many more, make up a sea of potential truths which corresponds to the sea of potential states of a quantum particle. These potential truths are not all mutually exclusive. It is entirely possible that the incident was planned media manipulation, and that Putin was also drunk. Many of the thoughts that follow the event are wrong, but there is no reason to think that only one is true.

There may be a lot of potential truths in that web of thoughts, but this does not mean that there is an infinite number. Vladimir Putin wrestling with a kangaroo would not generate the thought that carrots enjoy opera, for example. Events at the quantum scale may seem to be all over the shop, but it is not the case that anything goes.

The vast majority of these thoughts will not make it into the resulting media coverage. Russian journalists will self-censor because they believe it could be dangerous to express those thoughts in public, given Putin's attitude to restrictions on reporting and freedom of speech. In the West that self-censorship is more typically brought about by the influence of lawyers. Lawyers look at the content of the coverage and reject anything that they could not defend in a court of law. Conspiracy theories about media manipulation, mental breakdowns and alcohol abuse dissolve away under the cold hard legal gaze. A wild and enjoyable cloud of potential interpretations collapses down into something more prosaic and solid. Or at least, they do for most news outlets.

In our analogy the lawyers represent the gaze of scientists, who peep into their experiment to see what is going on. In the time between the news event occurring and the resulting media coverage, all sorts of wild and exciting thoughts are flying around. It is only the arrival of the lawyers, or the curious gaze of a scientist, that puts an end to the fun. It is this act of observing that causes a cloud of potential to solidify into a measurable result.

What would happen if Vladimir Putin released a statement after the fight, but before the newspaper headlines, in which he shed more light on the incident? Imagine that in this hypothetical statement Putin confessed to undergoing therapy to deal with his crippling phobia of marsupials. This would, at a stroke, change the face of the following morning's front pages. Headlines that would celebrate Putin's virile physicality would need to be rewritten. Anti-Putin headlines that questioned his suitability for office would be more extreme. Half of the possible results of the news event would wink out of existence before they even happened. Putin's official statement would, like the gaze of lawyers, shut down speculation and potential headlines. The coverage of a news event is, therefore, not only affected by legal intervention and interpretation, but it also varies depending on exactly when that intervention occurs.

The subatomic world, ultimately, is a fuzzy sea of guesswork and speculation which only commits to becoming clear and definite when observed. The exact nature and timing of that observation can change what that foam of maybes coalesces into. We can't directly see this fuzzy sea because our attempts to observe it cause it to solidify, just as we can't read the minds of journalists but only see the finished stories they produce. The quantum world is like the fun your teenage children and their friends have in their room. You know it exists because you can hear their shrieks and laughter throughout the house, but if you pop your head around the door, it immediately evaporates and leaves only a bunch of silent self-conscious adolescents. A parent cannot see this fun in much the same way that the sun cannot observe a shadow. And yet, it exists.

The story of quantum physics is the story of people failing to find adequate metaphors for reality. My use of Putin fighting a kangaroo may have seemed like desperate floundering, but it is relatively sane in comparison to some explanations of the quantum world. The most famous of these inadequate metaphors is Schrödinger's cat.

Schrödinger's cat is a thought experiment suggested by the Austrian physicist Erwin Schrödinger in 1935. It was intended to be a *reductio ad absurdum* attempt to highlight the inherent absurdity of the prevailing interpretation of quantum physics. But quantum physics is rather impervious to *reductio ad absurdum* attacks, for they are usually more accurate than sensible descriptions.

In Schrödinger's thought experiment, a cat is sealed in a box. The box also contains some equipment which may, or may not, kill the cat during a certain time period.

Quite why Schrödinger thought it would be a good idea to bring the killing of cats into this is not clear. The fact that he considered it a reasonable analogy could provide some insight into how desperate scientists were to find adequate descriptions for the subatomic world. Or he may just have been more of a dog person. Psychological questions about his choices aside, we are left with a cat that we are unable to see. With the box sealed, there is no way to know if the cat is alive or dead. The status of the cat, at this point, is that it is just as much dead as it is alive. The only way to resolve the status of the cat is to open the box and have a look. This, the thought experiment suggests, is how the quantum world works.

Where this thought experiment fails as a popular metaphor is that non-scientists intuitively think the cat in the box must always be either alive or dead, even if we don't know which until we take a look. The point that the metaphor is trying to get across is that the cat is both alive and dead at the same time. This unresolved contradiction is called a 'superposition'. A superposition refers to a particle simultaneously existing in all its theoretically possible states at the same time, in the same way that Twitter would contain all the differing and contradictory perspectives on Putin fighting a kangaroo, before an observer came along and caused those potential states to collapse into something definite or legally defensible.

Cats which are both dead and alive are impossible to imagine, even if you read a lot of Stephen King, so Schrödinger's thought experiment is flawed. It is trying to describe something that is difficult to imagine by saying that it is like something which is impossible to imagine. With metaphors like these, it is perhaps not surprising that quantum mechanics has a reputation for being incomprehensible.

The issue of metaphors and how much they influence our thinking was a problem that was being worked on in the 1920s by the Polish engineer and philosopher Alfred Korzybski. Korzybski was interested in how much of our understanding of the world was coloured by the structure of our languages, such as the troublesome verb 'to be'. In the grammar of a language such as English, it is not strange to say that one thing 'is' a different thing. We would think nothing of saying 'Bertrand Russell is clever,' for example, when the true situation we are trying to express is that 'Bertrand Russell appears clever to me.' By using words like 'is' we project our internal ideas, suspicions and prejudices onto the world around us, and then fool ourselves into thinking that they are externally real.

It is our ability to do this that allows us to lose ourselves in movies, and see fictional characters as real people rather than actors reading lines. But as Korzybski continually stressed, the map is not the territory. This was the point that the Belgian painter René Magritte was making in his 1929 painting *The Treachery of Images*, which depicts a smoker's pipe over the sentence '*Ceci n'est pas une pipe*' ('This is not a pipe'). This would initially baffle its audience until it was pointed out that the image was not an actual pipe but a picture of a pipe, at which point the message of the painting would become obvious.

The difficulty of distinguishing between metaphor and reality, the map and the territory, is one which is particularly problematic in quantum physics. The concept of multiple universes was first dreamt up 'after a slosh or two of sherry' by the American physicist Hugh Everett III in 1954. Everett was looking for a different interpretation of the prevailing understanding of quantum mechanics, and he came up with a real humdinger. What if instead of the cat being alive and dead at the same time, he thought, the cat was alive in our universe but dead in a completely different universe? If a vast number of parallel universes actually existed, then everything which can happen does happen. Every quantum superposition would be a list of the possible alternatives in other universes. The act of observing quantum events wouldn't cause clouds of potential to collapse into a single event, but would just remind us of what universe we are in.

The initial reaction to Everett's idea was not good. Physicists tend

to favour Occam's razor, the principle that when there are competing explanations, the simplest is more likely to be correct. Universes
are big things, and conjuring one out of thin air in order to make
sense of a living dead cat is quite a leap. As Everett's idea required
a phenomenal number of universes to ping into existence, he met
with a lot of negativity.

Everett attempted to explain his idea to the great quantum physicist Niels Bohr in 1959, but the meeting did not go well. Bohr's
colleague Léon Rosenfeld would later write that 'With regard to
Everett neither I nor even Niels Bohr could have any patience with
him, when he visited us in Copenhagen more than 12 years ago
in order to sell the hopelessly wrong ideas he had been encouraged, most unwisely, by Wheeler to develop. He was indescribably
stupid and could not understand the simplest things in quantum
mechanics.'

Discouraged, Everett quit theoretical physics and spent the rest
of his life as a defence analyst. He died suddenly in 1982, at the
age of fifty-one. As per his wishes his ashes were dumped in the
trash.

Although there was an increased interest in his ideas towards
the end of his life, he did not live long enough to see how seriously
they are now taken. There are many physicists who view the many-
worlds interpretation, as it is now known, as the best description
of reality that we have. According to the Oxford physicist David
Deutsch, 'The quantum theory of parallel universes is not some
troublesome, optional interpretation emerging from arcane theoretical considerations. It is the explanation – the only one that is
tenable – of a remarkable and counterintuitive reality.'

Deutsch's view is not universally shared by his colleagues. The
jury is still out on whether the many-worlds interpretation of
quantum mechanics is a true description of reality or whether it
is ridiculous. We are not yet able to say with any certainty whether
parallel universes exist, or whether scientists such as Everett and
Deutsch are simply falling for their own metaphors, and confusing
the map with the territory.

*

One of the most surprising discoveries about the atom was that it was mostly empty. To be more specific, an atom was 99.9999999999999 per cent nothing. If an atom is as big as St Paul's Cathedral, then the nucleus will be the size of a cricket ball and the electrons orbiting it will be like flies buzzing around inside an otherwise empty cathedral. Apart from the cricket ball and the flies, there is nothing else. If you take the entire human race and compress them, in order to remove all that empty space from their atoms, then the remaining matter will be about the size of a sugar cube. It is an odd thought, when you trust a chair to hold your weight, that the chair barely exists in any physical way. It is an even odder thought to realise that neither do you.

This is little comfort when you stub your toe on a doorframe. The world appears annoyingly solid in those circumstances. But there is a big difference between the world at the subatomic level and the human-scale world we live in. The chairs we sit on hold our weight, most of the time, and we are frustratingly unable to walk through walls. The bits of matter which do exist are locked in place by strong forces of attraction and repulsion, and these forces mean that nearly empty objects will refuse to pass through other nearly empty objects.

Why does the world appear so different to us? In the quantum world things bumble about in a vague state of potential existence and can happily be in a number of places at the same time. The human-scale world is not like this. A box of biscuits is never in two locations at once, sadly. Systems behave differently at different scales, and when one factor changes significantly an otherwise functioning system can collapse. At the human scale, the universe's vagueness and potential are removed.

The models physicists use to understand the quantum world are highly reliable at the scales they are designed for, but they are no use for larger things. They predict the likelihood of a particle existing by squaring the height of the associated wave, and in doing so neatly treat the quantum world as both waves and particles at the same time. These models allow us to build smaller and smaller microchips, among their many other uses. Yet they flatly contradict relativity, even though it perfectly describes the movements of big things like stars and planets.

The contradictions between relativity and quantum physics annoy the hell out of many physicists, who dream of finding one perfect 'Theory of Everything' which would accurately describe the universe. As yet, such a perfect multi-scale model remains elusive, and so our knowledge of the universe is dependent on which of our contradictory models we choose to use to understand it with. To gain a more objective understanding, we have to act like Picasso in his cubist period, and somehow merge multiple different viewpoints together into some strange-looking whole. What we know of the world depends on decisions that we make when we look at it. The dreaded poison of subjectivity is alive and well, and lurking at the borders of science.

One of the most unexpected aspects of the quantum world is known as Heisenberg's uncertainty principle. This was discovered in 1927 when the German physicist Werner Heisenberg proved that knowing a particle's momentum made it impossible to know its position, and vice versa. The more accurate we are about one of a complementary pair of variables, the less it is possible to know about the other. Digesting the implications of this was something of a dark night of the soul for many physicists.

More than anything, this was what was unsettling about the subatomic world. Sure, it's odd when one thing is in multiple places at the same time. It's weird that entangled particles can communicate instantaneously over vast distances. But it was Heisenberg's uncertainty principle that was as shocking to physicists as Gödel's Incompleteness Theorem had been to mathematicians. The problem was not that we didn't know the exact facts about physical reality, but that we couldn't know them.

The world of solid matter had turned out to be resting upon a bedrock as incomprehensible as the unconscious. Just as our conscious minds were only a bubble of rationality in a larger unconscious mind that we were usually unable to observe or understand, so the physical world of solid matter and comprehensible cause and effect was just a hiccup in a larger reality, produced by the quirk of a human-scale perspective. Mind and matter had both initially seemed understandable, but when we drilled down to find out what they rested on we found not firm foundations, but the unknowable and the incomprehensible. There was nothing down there that might conceivably come to act as our much needed new omphalos. Our world, both mind and matter, was a small bubble of coherence inside something so alien we haven't even been able to find adequate metaphors to describe it.

All this would take some coming to terms with. Fortunately, many writers and artists were prepared to try.

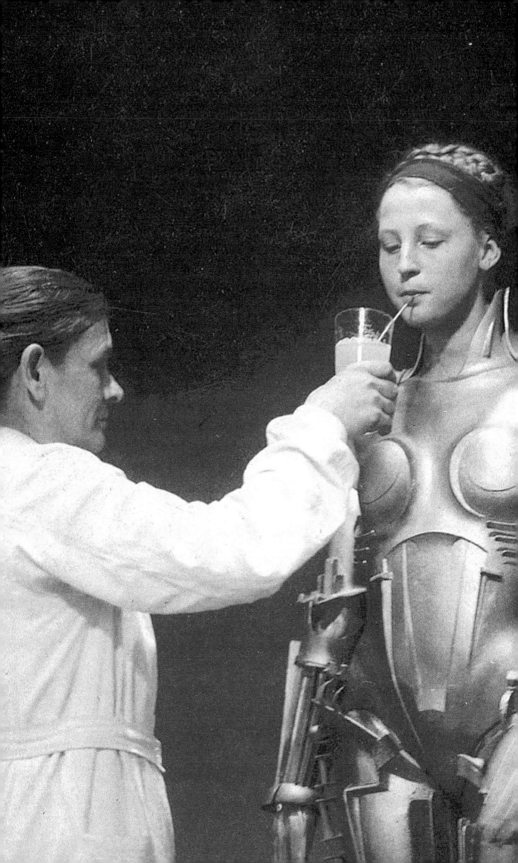

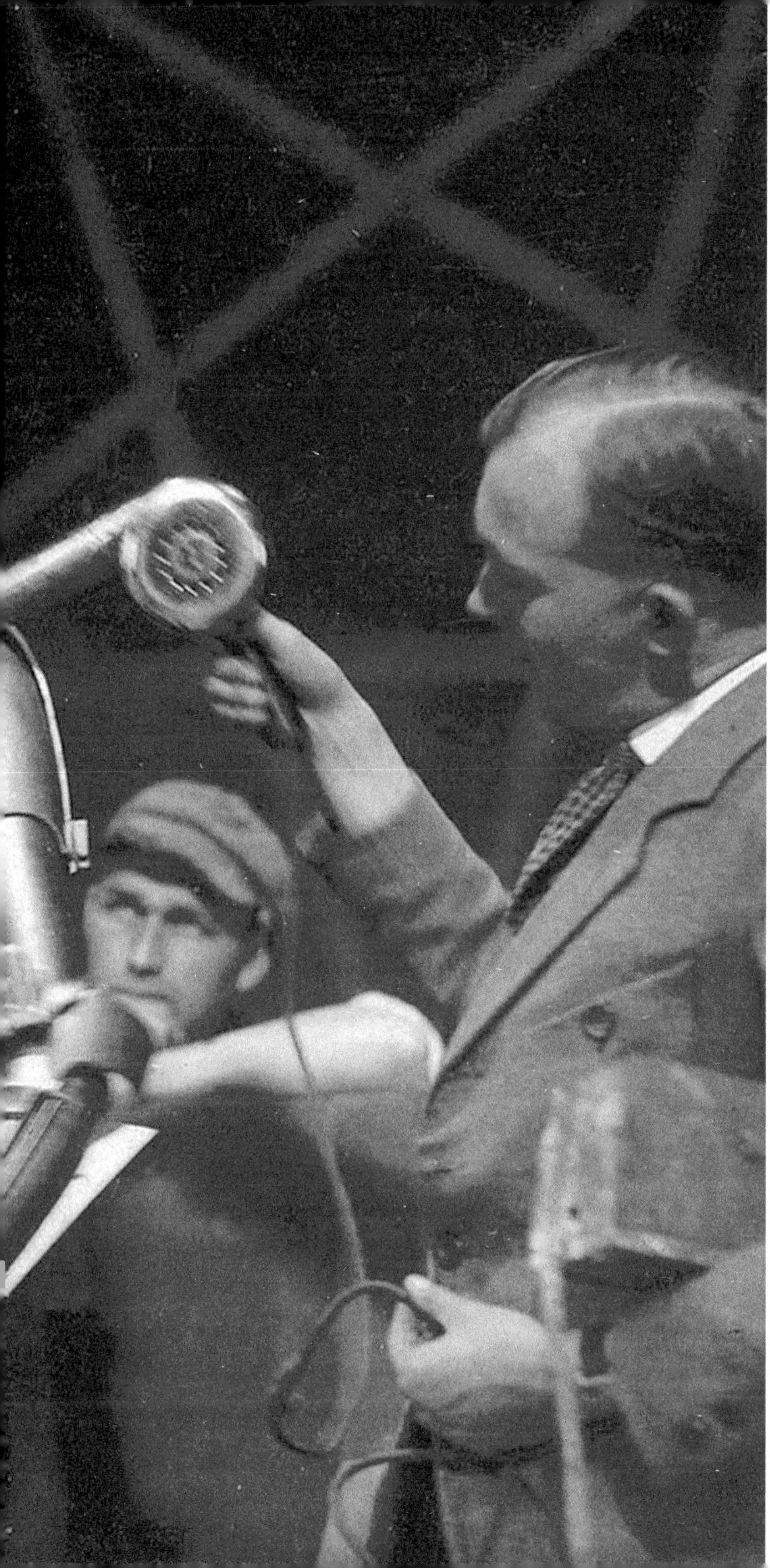

SEVEN: SCIENCE FICTION

Brigitte Helm, partially in costume, on the set of Metropolis, *1927* (Prismatic Pictures/Bridgeman)

A long time ago in a galaxy far, far away

The opening of the Chilean filmmaker Alejandro Jodorowsky's proposed adaptation of Frank Herbert's 1965 novel, *Dune*, would have been the most ambitious single shot in cinema.

It was to begin outside a spiral galaxy and then continuously track in, into the blazing light of billions of stars, past planets and wrecked spacecraft. The music was to be written and performed by Pink Floyd. The scene would have continued past convoys of mining trucks designed by the crème of European science fiction and surrealist artists, including Chris Foss, Moebius and H.R. Giger. We would see bands of space pirates attacking these craft and fighting to the death over their cargo, a life-giving drug known as Spice. Still the camera would continue forwards, past inhabited asteroids and the deep-space industrial complexes which refine the drug, until it found a small spacecraft carrying away the end result of this galactic economy: the dead bodies of those involved in the spice trade.

The shot would have been a couple of minutes long and would have established an entire universe. It was a wildly ambitious undertaking, especially in the pre-computer graphics days of cinema. But that wasn't going to deter Jodorowsky.

This scale of Jodorowsky's vision was a reflection of his philosophy of filmmaking. 'What is the goal of life? It is to create yourself a soul. For me, movies are an art more than an industry. The search for the human soul as painting, as literature, as poetry: movies are that for me,' he said. From that perspective, there was no point in settling for anything small. 'My ambition for *Dune* was for the film to be a Prophet, to change the young minds of all the world. For me *Dune* would be the coming of a God, an artistic and cinematic God.

For me the aim was not to make a picture, it was something deeper. I wanted to make something sacred.'

The English maverick theatre director Ken Campbell, who formed the Science Fiction Theatre Company of Liverpool in 1976, also recognised that this level of ambition could arise from science fiction, even if he viewed it from a more grounded perspective. 'When you think about it,' he explained, 'the entire history of literature is nothing more than people coming in and out of doors. Science fiction is about everything else.'

Jodorowsky began putting a team together, one capable of realising his dream. He chose collaborators he believed to be 'spiritual warriors'. The entire project seemed blessed by good fortune and synchronicity. When he decided that he needed superstars like Orson Welles, Salvador Dalí or Mick Jagger to play certain parts, he would somehow meet these people by happenstance and persuade them to agree. But when pre-production was complete, he went to pitch the film to the Hollywood studios.

Jodorowsky pitched *Dune* in the years before the success of *Star Wars*, when science fiction was still seen as strange and embarrassing. As impressive and groundbreaking as his pitch was, it was still a science fiction film. They all said 'no'.

By the time this genre was first named, in the 1920s, it was already marginalised. It was fine for the kids, needless to say, but critics looked down upon it. In many ways this was a blessing. Away from the cultural centre, science fiction authors were free to explore and experiment. In this less pressured environment science fiction became, in the opinion of the English novelist J.G. Ballard, the last genre capable of adequately representing present-day reality. Science fiction was able to get under the skin of the times in a different way to more respected literature. A century of uncertainty, relative perspectives and endless technological revolutions was frequently invisible to mainstream culture, but was not ignored by science fiction.

One example of how science fiction ideas could help us understand the psyche of the twentieth century was the Swiss psychoanalyst

Carl Jung's interest in UFOs. Jung, a protégé of Freud who famously split with his mentor over what he saw as Freud's sexual obsessions, wrote about flying saucers in 1959. He was eighty-three years old at the time, and hence not overly troubled by the effect this book would have on his scientific reputation.

UFOs were a phenomenon of the post-Second World War world. They arrived in mainstream consciousness following excited 1947 press reports of a UFO sighting by a credible American aviator called Kenneth Arnold. Right from the start they were assumed to be alien spacecraft from far-off worlds. Arnold had witnessed nine unidentified objects in the skies above Washington State. These objects were flat and half-moon-shaped, and looked something like a boomerang crossed with a croissant. He described them as moving like a fish flipping in the sun, or like a saucer skipping across water. From this, the press coined the phrase 'flying saucers'. Hundreds of further sightings followed, although interestingly these were of the circular, saucer-shaped objects suggested by the newspaper headlines rather than the bulging half-moon-shaped objects Arnold originally reported. After the press coined the term, witnesses reported that they too saw 'flying saucers'.

UFO reports evolved over the years. They were intrinsically linked to their portrayal in popular media, and sightings increased when movies such as *Close Encounters of the Third Kind* (1977) were released. Early accounts included encounters with aliens from Mars or Venus, whereas later reports, after those planets had been discovered to be lifeless, would tell of visitors from distant galaxies. New details became popular, such as abductions, cattle mutilations, big-eyed 'grey' aliens and, unexpectedly, anal probes. Links between aliens and secret military airbases became a big theme, although the fact that unusual craft were spotted above places used to develop secret military aircraft should not appear that strange.

Interest in the UFO phenomenon only really lost steam after the mass uptake of camera-equipped smartphones in the twenty-first century failed to produce convincing evidence of their existence. During their heyday, there was always something about UFOs that was more than just a story of nuts-and-bolts physical space travel. As the slogan of the 1990s television series *The X-Files* put it, 'I want to believe.'

Jung was not interested in the question of whether UFOs were 'real' or not. He wanted to know what their sudden appearance said about the late twentieth century. Mankind had always reported encounters with unexplained *somethings*, strange entities which, if they existed at all, were beyond our understanding. Jung understood that the interpretation of these encounters depended on the culture of the observer. Whether a witness reported meeting fairies, angels, demons or gods depended on which of those labels their culture found most plausible. The fact that people were now interpreting the 'other' in a new way suggested to Jung that a change had occurred in our collective unconscious.

As recently as the First World War, we still told stories about encounters with angels. One example, popularised by the Welsh author Arthur Machen, involved angels protecting the British Expeditionary Force at the Battle of Mons. But by the Second World War, Christianity had collapsed to the point where meetings with angels were no longer credible, and none of the previous labels for otherworldly entities seemed believable. As a result, the strange encounters which still occurred were now interpreted as contact with visitors from other planets. The ideas of science fiction were the best metaphor we had to make sense of what we didn't understand.

UFOs, to Jung, were a projection of Cold War paranoia and the alien nature of our technological progress. He recognised that the phenomenon told us more about our own culture than it did about alien spaceships. As he wrote, 'The projection-creating fantasy soars beyond the realm of earthly organizations and powers into the heavens, into interstellar space, where the rulers of human fate, the gods, once had their abode in the planets.' We no longer considered the heavens to be the domain of loving gods and angels.

The first science fiction story is often said to be Mary Shelley's *Frankenstein* (1818), because it was a story about a monster that was

neither natural nor paranormal, but which had been explicitly created in a laboratory. *Frankenstein* was about the desire to play God
and the discovery that this robs you of your humanity. It is perhaps no surprise that Shelley's story was so popular in the twentieth
century.

More typical examples of early science fiction arrive in the later
part of the nineteenth century. The *Voyages Extraordinaires* series
of adventure stories by the French writer Jules Verne included fantastic machines such as Captain Nemo's magnificent submarine the
Nautilus. Such technology, the stories warned, would be dangerous
in the wrong hands. The turn-of-the-century novels by H.G. Wells
also warned of the dangers of technology becoming a catalyst for
overreaching ambition, from the unnatural biological experiments
in *The Island of Doctor Moreau* (1896) to the tragic fate of *The Invisible Man* (1897).

Wells's work pioneered the use of science fiction to critique
current social problems. *The War of the Worlds* (1898) depicted
England invaded by a race of beings, in this case Martians, who were
determined, cruel and who possessed technology which was greatly
superior to that of the invaded natives. Wells found that he could
use this imaginative type of speculative fiction to present England
with a new and disturbing perspective on its own colonial history. Even the ending of the story, in which the advanced invaders
were stopped by native diseases, echoed the experiences of the British Empire. Social criticism was also evident in his 1895 novel *The
Time Machine*, which explored the implications of social inequality.
He projected the separation of the underclass and the privileged
elite into a nightmarish far future, where they had evolved into two
separate species, each horrific in its own way. Science fiction may
have talked about the future, but its power lay in what it said about
the present.

Late nineteenth-century science fiction may have been European and troubled by the implications of technology, but early
twentieth-century science fiction was very different. Science fiction
became American, and optimistic. Future technology was no longer

the harbinger of nightmares, but something positive and exciting. While Europe collapsed into industrialised warfare and introduced chemical weapons, tanks and aerial bombardment to world history, Americans dreamt about potential technology and found it all hugely thrilling.

This attitude is evident in the forty *Tom Swift* books, written by various ghostwriters under the pseudonym Victor Appleton between 1910 and 1941. Tom Swift, the son of an industrialist, is a gifted mechanic and inventor with a spirit of adventure and a plucky 'can-do' attitude. He is, as the English comics writer Alan Moore has noted, 'carefully depicted as a healthy macho male who's handy with his fists and whose scientific genius is mostly innate and self-taught without the need for any sissy book-learnin''. Each *Tom Swift* book details his adventures with a particular machine. They started out realistically, in books such as *Tom Swift and His Motor Cycle* and *Tom Swift and His Motor Boat* (both 1910), but quickly became more imaginative, such as the later titles *Tom Swift and His Sky Train* (1931) or *Tom Swift and His Magnetic Silencer* (1941). Some of these inventions were prophetic, including his invention of a proto-fax machine in *Tom Swift and His Photo Telephone* (1914). Modern tasers were inspired by *Tom Swift and His Electric Rifle* (1911), their trademarked name being an acronym of Thomas A. Swift's Electric Rifle.

This adventurous and optimistic take on invented technology quickly became the overriding attitude of early twentieth-century science fiction. Film and comic-book characters such as Buck Rogers and Flash Gordon were no-nonsense, all-American heroes who achieved their goals through action, bravery and whatever advanced technology happened to be lying around. They espoused individualism and held out the promise of an exciting future, as long as enough clear-sighted individuals did the right thing and built it.

These idealised attributes were not unique to stories of the future. They were also present in the cowboy stories which were at the height of their popularity during the same period. The romanticised version of the American frontier story began around the 1880s, when Buffalo Bill's Wild West show made repeated tours of Europe to great acclaim, and the dime novels of writers like Ned Buntline helped mythologise the lives of men like Buffalo Bill and Wild Bill Hickok. This was, noticeably, the period immediately after the frontier was tamed. The way of life that Buffalo Bill's show depicted was being replaced by the advance of law and civilisation, yet the myth of the Old West was becoming increasingly attractive, both in the US and beyond.

Beneath the genre trappings, Westerns and early science fiction stories were remarkably similar. *Star Trek*'s creator Gene Roddenberry famously sold his series by describing it as '*Wagon Train* to the stars', a reference to the long-running TV cowboy series. Secretly, he had more ambitious and progressive plans for it. 'Westerns were big and I wanted to sell [*Star Trek* to the network],' Roddenberry said in 1988. 'I said, "Look fellas, it's little more than a western. They have spaceships instead of horses and zap-guns instead of six-shooters, but it'll be familiar."' And unfortunately, they gave me the money and a set of good actors and a director, and I just went ape! They didn't get what they asked for or what we'd agreed on. They were naturally very upset.' Roddenberry's progressive aims were apparent when he threatened to walk away from the show unless Nichelle Nichols was cast in the role of Communications Officer Uhura, at a time when a black actress would not be cast in the role of a significant authority figure on American television. But while Roddenberry had a vision of the future above and beyond that described in his initial pitch, *Star Trek* was still a show about exploration and the frontier. America's mythologised past was powerful enough to also work as a mythologised future.

Many parts of the world had functionally equivalent versions of 'cowboys', from the *llaneros* on the Colombian Plains to the horsemen of Andalusia or the ranchers of the Australian outback. But these didn't grab the imagination in quite the same way as American cowboys. There was a magic ingredient in the American story that elevated it above stories of those similar lifestyles.

The myth of the Old West was a celebration of the power of the individual. It was a life free from state authority, where men had no lords or masters. People were considered equal and followed a clear moral code that prioritised reputation over wealth. The existence of a separate Native American population fed into this. Individualism requires a gulf between the self and the 'other', which the cultural differences between the existing and the colonising populations helped emphasise.

The British Marxist historian Eric Hobsbawm has highlighted the striking difference between the myths of the American and Canadian frontiers: 'One is a myth of a Hobbesian state of nature mitigated only by individual and collective self-help: licensed gunmen, posses of vigilantes and occasional cavalry charges. The other is the myth of the imposition of government and public order as symbolised by the uniforms of the Canadian version of the horseman-hero, the Royal Canadian Mounted Police.'

The logical end point of the idealised individualism of the Westerns genre was The Man With No Name, a character portrayed by Clint Eastwood in three films directed by Sergio Leone in the 1960s. This character was admired by the audience because he was so isolated and unattached to the wider community that he didn't even require a name. Like so many twentieth-century icons, his isolation was the cornerstone of his appeal.

A nation formed under the dream of a utopian future did not restrict its national myth to stories of its past. Not only did it invent new stories, it also mastered new mediums to tell them with.

Film had been born in the late nineteenth century. There wasn't a single inventor who could claim responsibility, but instead an accumulation of breakthroughs from inventors working independently around the world in working-class locations such as Brighton and Leeds in England, Lyon in France or New Jersey in the USA. By the start of the twentieth century filmmakers had learnt that they could cut between different shots without confusing the audience and were starting to experiment with techniques such as focus pulls and close-ups, but it still wasn't clear what direction the fledgling medium was going to develop in.

Cinema ultimately became a populist medium, arguably the most popular of the twentieth century, democratically supported and shaped by means of ticket sales. But it could easily have developed into a more elitist, highbrow art form. The great advances made by Russian, French and Italian filmmakers in the early decades of the century seemed to point in that direction. A film such as Giovanni Pastrone's *Cabiria* (Italy, 1914) towered over most films of the period, in both ambition and technical expertise. *Cabiria* depicted the destruction of the Roman fleet at the siege of Syracuse, the eruption of Mount Etna and Hannibal's trek across the Alps with elephants. It took six months to shoot, at a time when most films were completed in a few days. The monumental sets and huge crowd scenes still impress a hundred years later. In contrast, American cinema found initial commercial success by filming boxing matches, such as 1897's *The Corbett–Fitzsimmons Fight*, which remained a mainstream form of entertainment in the eyes of American audiences from that point on.

A number of events conspired to shift the centre of the film world away from Europe and towards California. The First World War destroyed much of Europe along with its economy and industry. The development of 'talking pictures' in the late 1920s gave the English-speaking film industry access to a larger world market than that of the French or the Italians. A third factor was the growth of Hollywood itself. It was a location which offered the reliable sunlight the industry needed and a relaxed, idyllic lifestyle that appealed to movie stars, which the industry increasingly depended on.

The American movie industry's move from the East Coast to California was in part promoted by patent disputes. The intention was to get as far away as possible from the lawyers of the MPPC, or Motion Picture Patents Company. The MPPC claimed intellectual ownership of the sprocket holes which mark the edges of each roll of film, and wanted to be handsomely rewarded. Considering the twenty-first-century industry's love of intellectual property, it is an irony that Hollywood itself was founded in the spirit of intellectual piracy.

From the very start, cinema and science fiction complemented each other wonderfully. Filmmakers knew that they needed to offer something beyond that which could be experienced by a trip to the theatre, so their ability to conjure fabulous visual special effects was clearly something they should exploit. Perhaps the most famous silent film is *A Trip to the Moon*, which was made by the French conjuror, showman and theatre impresario Georges Méliès in 1902. The film tells the story of six adventurous astronomers who build a rocket and fly to the moon. The team is led by Professor Barben-fouillis, which translates as Professor Messybeard. On the moon the adventurers are wowed by many wonders, including an appearance from the Moon goddess Selene, before being attacked by a race of alien insects and narrowly escaping back to earth.

The image of Professor Messybeard's rocket ship embedded in the eye of the Man in the Moon is only one of the many extraordinary visual images in that seventeen-minute film, but thanks to its simplicity, originality and humour it has become an icon of early cinema. *A Trip to the Moon* is considered so important to cinema history that it was the first film to be classified as a UNESCO World Heritage Film. It was clear from the very start, then, that cinema and science fiction would have a fruitful relationship.

The German expressionist masterpiece *Metropolis*, directed by Fritz Lang in 1927, is another example of this perfect marriage between medium and message. At its heart *Metropolis* was a loose version of the *Frankenstein* story. Much of the visual imagery we associate with *Frankenstein*, particularly the mad scientist's laboratory with its crackling electricity and strange machines, originates in this film and not in Mary Shelley's book. Its most famous scene depicts a lifeless female robot, built by a broken-hearted inventor named Rotwang, being given life as an evil double of the kind-hearted heroine Maria.

Metropolis is a film packed with insights into the psyche of the late 1920s. The visual design of this future city is a wild and uninhibited expression of modernist architecture. The unease about the dehumanising effect of industrialisation is apparent in the scenes of workers toiling away inside giant machines. The gulf between the drudgery of the workers and the luxurious lives of their rulers, at a time when labour-led revolutions still seemed a real possibility, is one of the key threads in the plot. Equally significant is the libidinous character of the female robot, which seduces the citizens of Metropolis into decadence and hedonism. As Gene Roddenberry would later realise, it was possible to use science fiction to talk about contemporary issues with a frankness that would have been unacceptable in more realistic narratives. All you had to do was distract the censor with the odd flying car.

The ambition of *Metropolis* was something of an anomaly in the story of early science fiction film. Early American science fiction cinema was more typical. This was concerned with populist adventure rather than bold artistic statements. It was typified by the films of Buster Crabbe, the Olympic Gold-winning swimmer who went on to play the roles of both Flash Gordon and Buck Rogers. These films did not portray the future as a utopia, exactly, but it still seemed an exciting destination.

After the Second World War science fiction films lost their youthful optimism. The fear of communism was apparent in B-movies such as 1956's *Invasion of the Body Snatchers*, a thriller in which the alien 'other' looked identical to everyday Americans. This was a film complex enough to be read as a condemnation of both communism and the paranoia of the anti-communist Senator Joseph McCarthy. *Soylent Green* (1973) presented a vision of global warming, overpopulation and a resource-short world, which echoed the growing environmental concerns of the age. The *Godzilla* series of Japanese monster movies reflected the Japanese relationship with nuclear technology. Godzilla was originally a monster who flattened cities but, in the years after Hiroshima and before Fukushima, he gradually evolved into a friend and protector of the Japanese people. *The Matrix* (1999), meanwhile, portrayed humanity as trapped in a virtual world and subservient to the computer technology it created.

In the second half of the twentieth century the adventure and excitement that Buster Crabbe represented were replaced by a growing unease. The genre now depicted harsh dystopias that individualism was unable to prevent. This did not affect the growth of individualism in the real world. On the contrary, its rise would continue onwards until the end of the century. The self-focused politics of Margaret Thatcher in Britain, who became prime minister in 1979, and her influence on President Ronald Reagan in the USA, would eventually make individualism the default political and economic perspective in the Anglo-Saxon world. But science fiction responds to subtler signals than politicians do. Like a canary in a coal mine, it is an early-warning signal. And postwar science fiction did seem to be warning us about something.

The script for George Lucas's 1977 movie *Star Wars* was influenced by *The Hero with a Thousand Faces*, a 1949 book by the American mythologist Joseph Campbell. Campbell believed that at the heart of all the wild and varied myths and stories which mankind has dreamt lies one single archetypal story of profound psychological importance. He called this the *monomyth*. As Campbell saw it, the myths and legends of the world were all imperfect variations on this one, pure story structure. As Campbell summarised the monomyth, 'A hero ventures forth from the world of common day into a region of supernatural wonder: fabulous forces are there encountered and a decisive victory is won: the hero comes back from this mysterious adventure with the power to bestow boons on his fellow man.'

Campbell found echoes of this story wherever he looked; myths as diverse as those of Ulysses, Osiris or Prometheus, the lives of religious figures such as Moses, Christ or Buddha, and in plays and stories ranging from Ancient Greece to Shakespeare and Dickens. This story is now known as 'The Hero's Journey'. It is a story that begins with an ordinary man (it is almost always a man) in a recognisable world. That man typically receives a call to adventure, encounters an older patriarchal mentor, undergoes many trials in his journey to confront and destroy a great evil, and returns to his previous life rewarded and transformed. George Lucas was always open about the fact that he consciously shaped the original *Star Wars* film into a modern expression of Campbell's monomyth, and has done much to raise the profile of Campbell and his work.

Star Wars was so successful that the American film industry has never really recovered. Together with the films of Lucas's friend Steven Spielberg, it changed Hollywood into an industry of blockbusters, tent-pole releases and high-concept pitches. American film was always a democratic affair which gave the audience what it wanted, and the audience demonstrated what they wanted by the purchase of tickets. The shock with which Hollywood reacted to *Star Wars* was as much about recognising how out of step with the audience's interests it had become as it was about how much money was up for grabs. Had it realised this a few years earlier, it might have green-lit Jodorowsky's *Dune*.

The fact that Lucas had used Campbell's monomyth as his tool for bottling magic did not go unnoticed. As far as Hollywood was concerned, The Hero's Journey was the goose that laid the golden eggs. Studio script-readers used it to analyse submitted scripts and determine whether or not they should be rejected. Screenwriting theorists and professionals internalised it, until they were unable to produce stories that differed from its basic structure. Readers and writers alike all knew at exactly which point in the script the hero needed their inciting incident, their reversal into their darkest hour and their third-act resolution. In an industry dominated by the bottom line and massive job insecurity, Campbell's monomyth gained a stranglehold over the structure of cinema.

Campbell's monomyth has been criticised for being Eurocentric and patriarchal. But it has a more significant problem, in that Campbell was wrong. There is not one pure archetypal story at the heart of human storytelling. The monomyth was not a treasure he discovered at the heart of myth, but an invention of his own that he projected onto the stories of the ages. It's unarguably a good story, but it is most definitely not the only one we have. As the American media critic Philip Sandifer notes, Campbell 'identified one story he liked about death and resurrection and proceeded to find every instance of it he could in world mythology. Having discovered a vast expanse of nails for his newfound hammer he declared that it was a fundamental aspect of human existence, ignoring the fact that there were a thousand other "fundamental stories" that you could also find in world mythology.'

Campbell's story revolves around one single individual, a lowly born person with whom the audience identifies. This hero is the single most important person in the world of the story, a fact understood not just by the hero, but by everyone else in that world. A triumph is only a triumph if the hero is responsible, and a tragedy is only a tragedy if it affects the hero personally. Supporting characters cheer or weep for the hero in ways they do not for other people. The death of a character the hero did not know is presented in a manner emotionally far removed from the death of someone the hero loved. Clearly, this was a story structure ideally suited to the prevailing culture. Out of all the potential monomyths that he could have run with, Campbell, a twentieth-century American, chose perhaps the most individualistic one possible.

The success of this monomyth in the later decades of the twentieth century is an indication of how firmly entrenched individualism became. Yet in the early twenty-first century, there are signs that this magic formula may be waning. The truly absorbing and successful narratives of our age are moving beyond the limited, individual perspective of The Hero's Journey. Critically applauded series like *The Wire* and mainstream commercial hit series such as *Game of Thrones* are loved for the complexity of their politics and group relationships. These are stories told not from the point of view of one person, but from many interrelated perspectives, and the relationships between a complex network of different characters can engage us more than the story of a single man being brave.

In the twenty-first century audiences are drawn to complicated, lengthy engagements with characters, from their own long-term avatar in World of Warcraft and other online gameworlds to characters like Doctor Who who have a fifty-years-plus history. The superhero films in the 'Marvel Cinematic Universe' are all connected, because Marvel understand that the sum is greater than the parts. A simple Hero's Journey story such as J.R.R. Tolkien's *The Hobbit* becomes, when adapted for a twenty-first-century cinema audience, a lengthy trilogy of films far more complex than the original book. We now seem to look for stories of greater complexity than can be offered by a single perspective.

If science fiction is our cultural early-warning system, its move away from individualism tells us something about the direction we are headed. This should grab our attention, especially when, in the years after the Second World War, it became apparent just how dark the cult of the self could get.

A still from Casablanca *featuring* (second from left to right) *Claude Rains, Paul Henreid, Humphrey Bogart and Ingrid Bergman, 1942* (Michael Ochs Archives/ Getty)

I stick my neck out for nobody

In the early 1940s a sign made from fifty-foot-high white letters stood in the Hollywood Hills. The letters read, 'OLLYWOOD-LAND'. They had originally been built in 1923 to advertise the Hollywoodland housing development. When it was new, light bulbs had lit up 'HOLLY', 'WOOD' and 'LAND' in sequence, and there was also a searchlight underneath it in case fifty-foot-high flashing letters were too subtle. But twenty years later the lights no longer worked and the sign was in need of repair, not least because a drunk-driver had left the road above and flattened the 'H' with his car.

The fate of the sign in the 1940s echoed the story of Hollywood, which also had a troubled start to the decade. Japan's attack on Pearl Harbor in 1941 and America's entry into the war closed off a number of foreign markets. Many skilled filmmakers quit the industry and enlisted, including the actors James Stewart and Clark Gable and the director Frank Capra. But Hollywood recovered from these set-backs and continued crafting the dreamtime of the Western world. By the end of the 1940s the sign had been rebuilt. In order to refer to the district, rather than the housing development, it had become the now globally recognised 'HOLLYWOOD'.

There was an urgency and boldness about the stories Holly-wood told in this decade. They were unavoidably coloured by the psychological impact of the Second World War. There is little of the sentimentality associated with Hollywood from the 1950s on-wards, and little of the whimsy that could be found in the silent era. Instead, stories were marked by a strong sense of purpose. Holly-wood continued to produce escapist fantasies, but when filmmakers talked about love and loss, or betrayal or duty, they talked about them plainly and honestly, to an audience with real experience of these states.

Television might have been slowly making its way into American homes, but it could not match Tinseltown for magic. Hollywood offered swashbucklers like Errol Flynn and Tyrone Power and the glamour of Ingrid Bergman or Bette Davis. There was entertainment from Judy Garland and Bob Hope, and Bugs Bunny and Lassie for the children. Hollywood was a place associated with wealth and elegance, with a hedonic sleazy underside. It is no wonder it seduced the world.

Hollywood was also an industry. The 'Big Five' major studios were MGM, Twentieth Century Fox, RKO, Paramount and Warner Bros. Each of these owned production facilities, theatre chains, distribution divisions and even the stars themselves. Each studio produced about a film a week. This mix of talent, experience and opportunity meant that the studio system could occasionally rise above art, and produce magic.

Casablanca (1942) is the story of Rick Blaine, a bitter exiled American. Rick owned a nightclub in Morocco which was bright, spacious and stylish, but full of desperate or corrupt clientele. He had to decide if his love for a beautiful former girlfriend was more important than helping an influential Resistance leader escape from the Nazis. It was a story with everything thrown into the mix – love, duty, patriotism, humour, romance, danger, friendship and desire. Each of these ingredients was heightened to the degree that only 1940s Hollywood could manage.

The heart of the film lies in Humphrey Bogart's career-defining performance as Rick. He plays a broken man, albeit a heavily glamorised one. He is a cynical and isolated anti-hero who repeatedly declares that he sticks his neck out for nobody. He gives his nationality as 'drunkard' and only drinks alone. It seems incredible now that the press release announcing the film's production stated that the role would be played by Ronald Reagan, who was then a popular actor.

The importance of the nihilistic side of Rick to the strength of the story was apparent from the beginning. The producer Hal Wallis sent an unproduced play called *Everybody Comes to Rick's* to a number of writers and filmmakers to gauge their opinion, including the Scottish screenwriter Æneas MacKenzie. Although the play would need extensive work in order to pass the strict moral rules of the censor, MacKenzie recognised 'an excellent theme' in the story that focused on the character of Rick. As he explained to Wallis, 'When people lose faith in their ideals, they are beaten before they begin to fight. That's what happened to France [in 1940], and it happened to Rick Blaine.' The final screenplay kept the idea that the character of Rick was a political allegory, but it aligned him more with post-Pearl Harbor America. When Rick tells the Chief of Police that 'I stick my neck out for nobody' he is told that this is 'a wise foreign policy', but the crime boss in charge of Casablanca's black market asks him bluntly, 'My dear Rick, when will you realise that, in this world, isolationism is no longer a practical policy?'

For many people in the middle of the twentieth century, the cynical, faithless character of Rick was someone they could relate to all too well. Anti-heroes with nothing to believe in became increasingly common as the 1940s rolled into the 1950s, as the certainty of purpose, which had characterised the war for the Allied nations, receded into memory. They were particularly prevalent in the works of the Beat writers, such as Scotland's Alexander Trocchi.

The central character in Trocchi's 1954 novel *Young Adam* is a young labourer called Joe Taylor, who worked on a barge on the canals between Glasgow and Edinburgh. Taylor is present when the body of a young woman in a petticoat is fished out of the canal. He keeps quiet about the fact that he knew the girl, that he had seen her trip and fall into the water, and that he did not attempt to save her. He also makes no attempt to help with the investigation into her death. When an innocent man called Daniel Gordon is tried for her murder, Taylor watches the trial from the public gallery. Gordon is found guilty and sentenced to death by hanging, but Taylor does not intervene to save him.

As Taylor sees it, why should he get involved? He feels separate and removed from his fellow Scots. What was in it for him, apart from hassle and potential danger? The authorities had already shown a willingness to hang an innocent man for the woman's death, and if Taylor came forward he had no guarantee that they wouldn't accuse him of murder. Like *Casablanca's* Rick Blaine, Taylor would not 'stick his neck out' for another person, even an innocent one whom he had the power to save from the hangman's noose.

Trocchi was a heroin addict, and what he captured so acutely in this novel was the morality of the junkie subculture. The junkie is separated from wider society by both the illegality of their drug and its isolating effects. Heroin users are not concerned with contributing to the greater good. They see society as something to be worked in order to fund their habits. That outsider lifestyle, separated from the anchors of work, respect and family, carries a heavy emotional cost, but that cost dissolves away under the effects of the drug. While on heroin, the problems of existence evaporate. The individual user becomes a self-contained unit, at peace with both themselves and their separation from others. The heroin subculture is, in many ways, the ultimate expression of individualism. You can understand why Aleister Crowley was so fond of the drug.

Trocchi detailed the junkie lifestyle in his most famous novel, *Cain's Book* (1960), which was subject to an obscenity trial in Britain in 1965. This heavily autobiographical story, about an unrepentant homosexual junkie working on a scow on the Hudson River in New York State, describes the relationships between a number of addicts. In this subculture the primacy of each person's individual need for heroin is understood and the inevitable thefts and betrayals between them are accepted. Their relationships are those of convenience with no expectation of concern or kindness, yet rare moments of tenderness do somehow arise.

Trocchi saw this extreme individualism as a reflection of the wider culture and something that novelists had to confront. 'Writing which is not ostensibly self-conscious is in a vital way inauthentic for our time,' he wrote in *Cain's Book*. He was aware of how isolating such individualism was. As the novel's narrator admits, 'It occurred to me that I was alone. And then it occurred to me how often that thought occurred to me.' Yet he was insistent that isolated individualism was necessary, and repeatedly defended the junkie lifestyle. 'It is impertinent, insolent, and presumptuous of any person or group of persons to impose their unexamined moral prohibitions upon me, it is dangerous both to me and, although they are unaware of it, to the imposers,' he wrote. 'God knows there are enough natural limits to human knowledge without our suffering willingly those that are enforced upon us by an ignorantly rationalized fear of experience.'

Trocchi's unrepentant love of heroin, together with his wife Lyn Trocchi's willingness to prostitute herself to fund both their habits, probably turned more people away from extreme individualism than it converted. Andrew Scott Murray's 1992 biography of Trocchi is entitled *The Making of a Monster*. Yet Trocchi's belief that such a nihilist worldview needed to be faced and accepted came with heavy intellectual backing.

Jean-Paul Sartre was a small, round-faced man with even rounder glasses. He became the postwar icon of the French intelligentsia, typically found smoking in his favourite left-bank Parisian café with his partner, the writer Simone de Beauvoir. Sartre's fame came through the success of his first novel, *Nausea*.

The novel was first published in France in 1938. The Second World War broke out the following year and Sartre was drafted into the French army, despite being virtually blind in one eye. He found himself in a German prisoner-of-war camp along with a million and a half other French troops shortly afterwards. It was only after the war that the full extent of his book's impact became apparent.

When the troops returned home there was a palpable desire to move beyond the horrors of the conflict and for life to return to normality. But what, exactly, was normality? The scientific, artistic and industrial breakthroughs from between the wars had come so thick and fast that nobody had had time to come to terms with them. The political and moral crises caused by the emergence of fascism and communism had been a more pressing concern. It was not until peace arrived that people had time to take stock and re-evaluate where they were. What they found came as something of a shock. The great pre-First World War certainties were clearly no longer viable, but what, exactly, had replaced them? Uncertainty, paradox and irrationality were everywhere. Where was the postwar omphalos, the perspective from which everything made sense? Countless isolated individual perspectives all came to the realisation that there wasn't one.

Nausea found its audience in a world that had witnessed the use of nuclear weapons against the civilian populations of Hiroshima and Nagasaki. In that book Sartre wrote that 'Everything that exists is born for no reason, carries on living through weakness, and dies by accident.' The core principle of existentialism is the recognition that life is meaningless, and that the experience of existing in the present moment is all that matters. The rest of Sartre's philosophy was an attempt to come to terms with this.

Sartre didn't waste time arguing that there was no God. He understood that most thinking people of the time had already reached that conclusion themselves. At the point when Sartre's novel found its audience, in the immediate aftermath of the Holocaust, there were few who would point at the world and declare that it was the work of a just and moral god. Sartre's aim was to explore what it meant to be alive in a godless universe.

Nausea is the story of an isolated academic called Antoine Roquentin. Roquentin lives alone in a hotel room in a small French seaport called Bouville, or 'Mudtown', which Sartre based on Le Havre. He has a small private income which allows him to work on a biography of an obscure historical figure. The book takes the form of Roquentin's diary, in which he attempts to understand a change in his perception of the world. This shift in perception results in a feeling of sickness, or nausea, which is initially triggered by nothing more than becoming aware of a stone on the beach. Roquentin believes that he has seen past his mental concept of the stone, what he calls the 'essence' of things, and encountered the stone as it actually is. It has no purpose or reason to exist, and the fact that it does makes Roquentin nauseous.

Roquentin seeks an escape from the despair of meaninglessness. Many possible escape routes are considered, including education, adventure, inner contemplation, love, town life and the past, but all of these are found to be illusions compared to the sheer overbearing reality of existence. The book allows for the possibility that art offers an escape from existence, but this rare moment of optimism was something that Sartre would himself reject after the war.

For Sartre and for Trocchi, denying the meaninglessness of existence was cowardice. It was necessary to fully confront our situation because, as Sartre saw it, 'Life begins on the other side of despair.' He believed that we were both blessed with free will yet cursed with the awareness of the pointlessness of it all, which meant that mankind was 'condemned to be free'.

Sartre was an admirer of another leading nihilist, the Irish-born author and playwright Samuel Beckett. Sartre published the first half of a short story by Beckett in *Les Temps Modernes*, a magazine he edited, not realising that the story was incomplete. When the error was discovered, he did not see any point in publishing the second half.

Beckett's most famous work is his 1953 play *Waiting for Godot*. It is about two men, Vladimir and Estragon, who are dressed like vagrants but talk like academics, and who spend the entire play waiting for the arrival of a character known as Godot, who doesn't show up. It was, as the critic Vivian Mercier noted, 'a play in which nothing happens, yet that keeps audiences glued to their seats. What's more, since the second act is a subtly different reprise of the first, [Beckett] has written a play in which nothing happens, twice.'

The originality of *Godot* made Beckett famous, but the nihilism it espoused painted him into a corner. Once you've said that life is meaningless there is little more you can add. Sartre escaped this problem by attempting to build a grand synthesis between existentialism and Marxism, but Beckett was not interested in searching for a way out of nihilism. He continued to produce bleak and despairing work.

His follow-up to *Waiting for Godot* was called *Endgame*. The main character is Hamm, a blind cripple, whose legless parents Nagg and Nell are sitting nearby in dustbins. The final character is Hamm's adopted son Clov, who has stiff legs which do not allow him to sit down, and which force him to walk in a ridiculous manner. Nothing much happens, but that nothing is relentlessly bleaker and less entertaining than the nothing that happens in *Godot*. It was as if Beckett had decided to embrace self-parody before anyone else could parody him.

Nevertheless, the literary establishment were very taken by the rise of nihilism and the Nobel Prize for Literature was offered to many of its leading practitioners. Sartre turned the award down, which made ideological sense. Existentialism says that everything is meaningless, not that everything is meaningless except literary awards. The Algerian-born French writer Albert Camus, whose 1942 novel *The Stranger* was a key text in the growing movement, did accept the prize. In his defence he always claimed that he was an absurdist, not an existentialist. Camus believed that life was absurd, rather than meaningless. He considered that an important distinction.

The English writer Colin Wilson was struck by the difficulties *Endgame* presented to the critics. Wilson was then highly lauded, in the wake of the success of his 1956 book *The Outsider*. He was initially linked with a loose movement of disillusioned British writers known as the Angry Young Men, although his literary cachet fell when later books revealed he wasn't that angry or disillusioned. Wilson attended the London premiere of *Endgame*, having already seen the play in Paris. Clearly neither the audience nor the critics really liked the play, or so it seemed to him, but they were wary of criticising it because nihilism was so tricky to argue with. Those who dismissed it on the grounds that things just weren't that bad were told that they were in denial, or that they lacked the courage to look as deeply into the nature of things as Sartre, Beckett or Camus did.

But Wilson had, as we noted in Chapter 2, personally experienced the state that psychologists call flow, and which he called peak experience. As a result of this, he too felt that he had seen the true existential nature of things, and it did not produce nausea in him. From Wilson's perspective, life was wonderful and self-justifying, and the act of living was undeniably of value.

For Wilson, the existentialists were like people who walked round an art gallery in the dark and declared that there was nothing of interest there. His peak experience, in contrast, had been like switching on the light. It didn't cause the art in the gallery to suddenly appear, because it was always there. Instead, it made it undeniable. It seems possible that Camus also came to understand this. As he wrote in 1952, 'In the middle of winter I at last discovered that there was in me an invincible summer.'

Those who attempt to describe the flow state stress the loss of the sense of self. Achieving Wilson's insight required deep involvement with something external, which individualist philosophies actively fight against. The protagonists of existentialist novels were almost always passive, isolated figures, while many leading existentialist figures, such as Sartre and Beckett, were children of wealthy families who went through life without normal financial pressures. When Beckett was a young man he once stayed in bed all day, because he couldn't see any reason to get up. This apathy was the exact opposite of the engagement and interest in the world needed to trigger the flow state.

The passive navel-gazing of a nihilist was a self-fulfilling prophecy. If you engaged in it, then life did indeed appear meaningless. But that perspective, crucially, only described the individual nihilist and not mankind in general. Outside of the existentialist bubble, there was value and meaning to be found. It did not require intellectual gymnastics, or faith in God or Marx. It just required energy, dedication and a desire to get involved.

Wilson's disagreement with Beckett was philosophical rather

than creative, so unlike most critics he was not distracted by how funny, original or creative Beckett could be. Their philosophical disagreement was exemplified by Clov's final soliloquy, when he remarks that 'I say to myself that the earth is extinguished, though I never saw it lit.' 'This explains why,' Wilson has written, 'when I went to the first night of *Endgame*, I rejected it as an attempt to convince me that black is white. When Clov says that the world is going out, but he has never seen it lit up, I could say 'Well *I* have,' and dismiss Beckett as a man suffering from laziness and self-pity – both his own fault.'

In Sartre's *Nausea*, the protagonist Antoine Roquentin initially suspects that the repulsion he has begun to feel towards existence may not be a product of the objective world, but rather something internal that he projects outwards. 'So a change has taken place during these last few weeks', he says. 'But where? It is an abstract change without object. Am I the one who has changed? If not, then it is this room, this city and this nature; I must choose.' Roquentin initially declares that 'I think I'm the one who has changed: that's the simplest solution. Also the most unpleasant,' yet as he sinks deeper and deeper into despair he increasingly blames the world and confuses his internal impressions with the nature of the world outside him.

Both Beckett and Wilson made observations about the true nature of things. Yet because they had different perspectives, those observations were different. What was observed, once again, was dependent on the observer. As William Blake wrote at the start of the nineteenth century, 'For the eye altering, alters all.'

Existentialism also flowered in the United States, but American existentialists were considerably more engaged with life than their Irish and French counterparts.

Jack Kerouac was an athletic, Catholic-raised writer from Massachusetts. He gained a scholarship to Columbia University on the strength of his American football skills, but dropped out and gravitated to the bohemian underworld of New York City. This subculture inspired his book *On the Road*, the most famous of all the Beat novels.

On the Road was written during an intense three-week period in 1951. It was typed, single spaced and without paragraph breaks, onto a continuous 120-foot-long scroll of paper that had been made by taping together separate sheets of tracing paper. Kerouac hammered away at the typewriter for long hours at a stretch, fuelled by amphetamines and not stopping for food or sleep. The scroll meant that he did not need to stop in order to insert a new page. The result was a stream-of-consciousness outpouring of pure enthusiasm which had the rhythms of jazz. It was as if he were constantly ramping up the energy of his prose in order to outpace and escape the nihilism of the world he wrote about. Kerouac's writings are peppered with references to the Buddhist concept of *satori*, a mental state in which the individual perceives the true nature of things. The true nature of things, to someone experiencing satori, was very different to the true nature of things as perceived by Sartre or Beckett.

It was Kerouac who coined the phrase 'the Beat Generation'. The name arose in conversation with his friend John Clellon Holmes. As he later recalled, '[John] and I were sitting around trying to think up the meaning of the Lost Generation and the subsequent existentialism and I said "You know John, this is really a beat generation"; and he leapt up and said, "That's it, that's right!"'

Although many people in the underground drug culture of the 1940s and 1950s self-identified as both 'hipsters' and 'Beats', the term 'Beat' has since gained a more specific definition. The phrase 'Beat Generation' is now used mainly to refer to the American writers Jack Kerouac, Allen Ginsberg and their muse Neal Cassady. Original nihilistic Beat writers such as Trocchi are left out of this definition. There are also attempts to include the American writer William Burroughs in the Beat Generation, despite Burroughs's unique ability to escape from any category he is placed in. This narrowing of focus has led the American poet Gregory Corso to remark 'Three writers do not a generation make.'

Kerouac had originally picked up the word 'beat' from a street hustler and junkie who used the term to sum up the experience of having no money or prospects. Kerouac's imagination latched onto the word because he saw another aspect to it, and one which complemented its original meaning of referring to a societal outcast. For Kerouac, the word implied *beatitude*. Beatitude, in Kerouac's Catholic upbringing, was the state of being spiritually blessed. Shunned outcasts who gain glimpses of grace and rapture are a constant theme of Kerouac's work, and the word summed this up in one immediate single-syllable blast. The word became attached to the wild, vibrant music that the Beats were so attracted to, and it was this connection to 'beat music' that lies behind the name of The Beatles.

Unlike the more nihilistic European Beats, the American Beat Generation flavoured their version of existentialism with Eastern mysticism. By the middle of the twentieth century a number of Eastern texts had become available in Western translations. Richard Wilhelm's 1924 German translation of the *I Ching* was published in English in 1950, for example, and the American anthropologist Walter Evans-Wentz's 1929 translation of the *Bardo Thodol*, better known as the *Tibetan Book of the Dead*, gained widespread attention following its 1960 reissue. These texts described a spirituality that differed greatly from the hierarchical monotheistic religions of Judaism and Christianity, with their subservient devotion to a 'Lord'. They talked about divinity as being something internal rather than external. They viewed spirituality as a matter of individual awareness.

There are distinct differences in the definitions of words like satori, beatitude, enlightenment, grace, rapture, peak experience or flow, but these terms also have much in common. They all refer to a state of mind achievable in the here and now, rather than in a hypothetical future. They are all concerned with a loss of the ego and an awareness of a connection to something larger than the self. They all reveal the act of living to be self-evidently worthwhile. In this they stand in contrast to the current of individualism that coursed through the twentieth century, whose logical outcome was the isolation of the junkies and the nihilism of the existentialists.

But interest in these states, and indeed experience of them, were not widespread. They were the products of the counterculture and obscure corners of academia, and hence were treated with suspicion, if not hostility. The desire for personal freedom, which individualism had stoked, was not going to go away, especially in a generation that had sacrificed so much in the fight against fascism.

How could we maintain those freedoms, while avoiding the isolation and nihilism inherent in individualism? Reaching out towards satori or peak experience may have been one answer, but these states were frustratingly elusive and too difficult to achieve to provide a widespread solution.

The writers of *Casablanca* had difficulty finding the right ending for the film, but the script they turned in at the last minute created one of the great scenes in cinema. It takes place at Casablanca airport during a misty night, and includes a waiting plane, a dead Nazi, and a life-changing decision. Humphrey Bogart's Rick Blaine makes the decision not to leave Casablanca with Ilsa, the love of his life. He instead convinces her to leave with her husband and help him in his work for the Resistance. 'I'm no good at being noble,' he tells her, 'but it doesn't take much to see that the problems of three little people don't amount to a hill of beans in this crazy world.' This is the moment when he admits that there is something more important than his own individual perspective and desires. Although he previously declared that he stuck his neck out for nobody, he now risks his life and liberty in order to allow the Resistance leader to escape. Rick ended the film leaving for a Free France garrison so that he too could fight the good fight.

Hollywood movies fought off nihilism by offering hope, either through personal love, symbolic escape or the vaguely defined better future of the American Dream. Occasionally, they would offer warnings. Oscar-winning films such as *Citizen Kane*, *There Will Be Blood* or *The Aviator* were tragedies which depicted the ultimate isolation of those who got what they wanted.

Casablanca's screenwriters were helped by the fact that the film was set and made during the Second World War. This gave them a clear 'greater good' which they could appeal to. Rick was able to leave his spiritual and personal isolation in order to dedicate himself to the anti-fascist cause. But the film continued to resonate with audiences long after that war had been won, because Rick's escape from nihilism remained powerful on a symbolic level. The promise that there was something better than individual isolation was something that audiences craved. That something better, whatever it was, would take effort, and involvement. But that effort would make it worth working towards.

Existentialism lingered in Europe, but America was too industrious to navel-gaze. As the Second World War receded into memory, the United States was about to show the world exactly what mankind was capable of. It was time, President Kennedy boldly announced, to go to the moon.

The first test of a captured German V-2 rocket in the United States, 1946 (Universal History Archive/UIG/Getty)

We came in peace for all mankind

The moon has cast a spell on us since before we were human. After our nocturnal animal ancestors evolved the ability to focus on distant objects, they looked out from the treetops and saw that the moon was something *different*. It crossed the night sky in a manner unlike every other part of the natural world. It moved smoothly. It grew full and waned to a regular rhythm that was unaffected by events in the rest of the world, but which cast an inescapable influence over that world. It was separate from us, unreachable.

In time the moon became associated with dreams, love, longing and imagination, and all that was intangible. It was something that we yearned for, but could never claim. This didn't stop people fantasising about travelling there. The second-century Syrian writer Lucian claimed that a waterspout had transported his ship to the moon, in a book admirably entitled *A True History*. He found himself in the middle of a war between the King of the Moon and the King of the Sun. There were no women on Lucian's moon, and children had to be born from men. In the early seventeenth century the Welsh bishop Francis Godwin wrote of travelling to the moon in a vehicle pulled by wild swans. The moon was a utopian paradise, he reported, populated by lunar Christians. When Jules Verne wrote his novel *From the Earth to the Moon* in 1865, in which members of the Baltimore Gun Club built a giant cannon and fired themselves into space, it seemed just as fantastical as the stories of Godwin and Lucian.

In the 1960s the unreachable was claimed and the dream became reality. Going to the moon required romantic madness in order to believe that it could be done, and practical genius to make it happen. This is a rare and complicated psychological makeup, and one which has its dark side. Landing on the moon is still the single greatest achievement in history, and one dedicated to 'all mankind'. Yet it was also an act of single-minded determination and it took wild, dedicated individuals to achieve it. As the sociologist William Bainbridge observed, 'Not the public will, but private fanaticism drove men to the moon.'

The cosmos itself, as we understood it when Einstein's theory was first published, contained the planet Earth and seven other planets revolving around the sun. Pluto was not discovered until 1930. It was clear that, beyond our immediate solar system, there did seem to be an awful lot of other stars out there, but quite exactly what that meant was open to debate. This ignorance did not last, and the story of cosmology in the twentieth century was one of a continuous expansion of both our knowledge and our sense of awe.

In March 1919 the English astrophysicist Arthur Eddington sailed to the African island of Principe. His goal was to record the position of the stars during a solar eclipse, in order to find out if Einstein was right. Common sense and Newton's laws said that light from distant stars would be unaffected when it travelled close to the sun. But if the fabric of reality did curve in the presence of mass, then the path of light from distant stars would also curve. Those stars would appear to be in a slightly different position when the sun passed them. This could only be tested during a solar eclipse because the brightness of the sun made recording the position of the stars around it impossible.

Eddington's mission was a success. The universe behaved as Einstein's theory predicted. But Einstein's work predicted a lot of other strange things as well. It suggested the possibility of 'black holes', areas of matter so dense that everything nearby was pulled inescapably into them, including light. Relativity also claimed that the curved nature of space-time allowed the universe to be a finite size, but not come to an end. A spaceship travelling the length of the cosmos would not eventually fall off the end of the universe. Instead, it would find itself back where it started, like an ant walking around the circumference of a football wondering when the damn ball would come to an end. The concept of the centre of the universe, which made perfect sense in a three-dimensional universe, was meaningless in the four-dimensional universe of space-time. It was simply not possible to locate the edges of the universe, let alone work out a midway point between them.

Observations of the universe improved continually over the century. It became apparent that the air pollution over London hindered the work of the Royal Observatory in Greenwich, so the telescopes were packed away in 1948 and moved down to the clearer air of Herstmonceux in Sussex. Omphaloi may claim to be fixed points, but they never last. By 1984 Sussex was also deemed unsuitable, so the observatory was moved to the Canary Islands. High altitude and remote locations provided the greatest views of the heavens, so telescopes were built in locations in Chile, California and Hawaii. Even these began to show their limits, and some of our greatest telescopes are now placed above the atmosphere, in orbit around the planet. As a result the level of detail in our images of space increased exponentially during the twentieth century.

It gradually became apparent that the cosmos wasn't just sitting there, eternal and unchanging. The universe was expanding, like a balloon being inflated. And if the universe was expanding, then it stood to reason that it used to be smaller. If you went far enough back in time it would get smaller and smaller until it had shrunk down to nothing. This was the birth of the universe, the moment when the cosmos was born out of the void. In 1949 the English astronomer Fred Hoyle memorably described this as a 'Big Bang', although the event he described wasn't big and didn't go bang. This was, in theological terms, something of a game-changer. The universe was no longer 'just there', supporting us. It had been born, it was growing and one day, perhaps, it would die.

Our knowledge of the universe grew as our telescopes improved, and as a result mankind's relative importance grew smaller and smaller. The universe turned out to be full of clumps of stars called galaxies, such as our own local clump the Milky Way. These vary in size but can contain as many as a hundred trillion stars. There are believed to be more than 170 billion galaxies in the observable universe. Writing numbers like '170 billion' or 'a hundred trillion' is in many ways a pointless exercise, because those words in no way convey the quantity that they represent. Should a person even begin to glimpse what those figures represented, they would immediately need to sit down and have a strong drink. If they truly understood the scale of those trillions they would be off work for quite some time.

In the twentieth century we looked out into space, and discovered that we couldn't grasp how big it was without causing our minds to snap. This, then, was the frontier we were planning on crossing. This was the final frontier, a non-infinite infinity which could generate awe like nothing the human race had encountered before. It was time to leave home and take our first steps outside.

When he was a young boy growing up in pre-war California, Marvel Whiteside Parsons loved science fiction stories such as those found in Hugo Gernsback's *Amazing Stories* magazine. A particular favourite was Jules Verne's novel *From the Earth to the Moon*.

The idea that a rocket could leave the earth's atmosphere and travel to the moon was considered as fanciful then as a time machine is today. Rockets had existed for thousands of years, ever since the Chinese invented gunpowder, and their inclusion in 'The Star-Spangled Banner' ('. . . and the rocket's red glare') gave them a place in the American psyche. But they did not scale in a way which made journeys into space appear possible. The weight of the required fuel and the structural integrity needed to control such force seemed to be insurmountable obstacles. A 1931 textbook declared that there was 'no hope' that rockets would lead to space flight, and that 'only those who are unfamiliar with the physical factors involved believe that such adventures will ever pass beyond the realm of fancy.' As late as 1940 Dr John Stewart, the Associate Professor of Astronomical Physics at Princeton University, wrote that while a rocket trip to the moon wasn't theoretically impossible he didn't expect it to happen before 2050. He had no idea that, the previous October, Nazi rocket scientists had launched a rocket to an altitude of almost sixty miles, very close to the 62-mile-high Kármán line, which marks the boundary between earth's atmosphere and outer space.

Regardless of what the experts thought, young Marvel Parsons was going to build such a rocket. He knew that Captain Nemo's submarine *Nautilus* had seemed unbelievable when it first appeared in Verne's *Twenty Thousand Leagues under the Sea* (1869), and similar vessels had since become a reality. Over the course of his short life Parsons would experiment, invent and, through hard work and a dash of genius, pioneer the solid-fuel rocketry that would take America into space, most notably in the solid rocket boosters that launched the Space Shuttle. He also invented jet-assisted take-off (JATO), which was a significant help to the American war effort, and was a co-founder of the Jet Propulsion Laboratory and the Aerojet Corporation. In the opinion of his biographer John Carter, 'everything today in the field of solid fuel rockets is essentially Parsons' work, if slightly modified.'

But Parsons was a complicated individual. He signed a document stating that he was the Antichrist. He dedicated his spiritual life to summoning the Whore of Babylon, the lustful, beast-riding divinity prophesied in the Book of Revelation, in order that She could claim dominion over the entire world. Parsons was born on 2 October 1914, which happened to be the date that Charles Taze Russell, the founder of the Jehovah's Witnesses, claimed would usher in Armageddon.

He rejected the name Marvel, in favour of Jack or John, when he was still young. Marvel was the name of his absentee father, whom he had come to hate. Parsons wrote about his desire to 'exteriorize [his] Oedipus complex', and there are rumours that home movie footage existed of him having sex with his mother. And also with his mother's dog. The only person Parsons would call 'father' was

Aleister Crowley, who he both idolised and supported with money earned from his career as a rocket scientist. Parsons would chant Crowley's *Hymn to Pan* before rocket tests, slowly stamping along with the words:

Thrill with lissome lust of the light,
O man! My man!
Come careering out of the night
Of Pan! Io Pan!

Nowadays chanting black magic invocations before rocket tests is frowned upon, but it does add a certain something.

Parsons was recruited into the world of academic aeronautics research by the famed Hungarian physicist Theodore von Kármán, after whom the boundary between earth's atmosphere and outer space was named. Von Kármán had a reputation for being willing to take on unlikely projects, and his colleagues at Caltech were happy to leave the 'Buck Rogers stuff' to him. Parsons had no formal college education, but von Kármán recognised his talent and intelligence and included him in a research group working on identifying more powerful rocket fuels. Naturally charming and handsome, Parsons had no difficulty moving among the engineers and experimenters of academia.

Parsons's group soon earned the nickname the Suicide Squad, following a number of failed rocket-fuel experiments that caused safety concerns on the Caltech campus. In response they were moved to a few acres of land nearer the San Gabriel Mountains, just above the Devils Gate Dam. NASA's Jet Propulsion Laboratory is situated there to this day, and considers its official moment of founding to be the experiments performed by Parsons and the Suicide Squad on Hallowe'en 1936.

The approaching war brought a turnaround in the fortunes of the group, in terms of both financial support and the credibility of their field. The outbreak of war in Europe also coincided with Parsons's discovery of Aleister Crowley, although he had long held an interest in the darker side of the occult. He claimed that he first attempted to invoke Satan at the age of thirteen. After the war, when his technical reputation was assured, he sold his share in the Eurojet Corporation and dedicated himself to furthering his occult studies.

Rumours began to circulate about the ungodly activities occurring at his large house on Pasadena's 'millionaires' row'. His home became a focus for both devotees of the occult and Los Angeles science fiction enthusiasts. His well-heeled neighbours were not happy when he began renting out rooms to 'undesirables', such as bohemians, artists or anarchists. Parsons had placed an advert in the local paper's 'rooms to let' section which advised that prospective tenants 'must not believe in God'. His bedroom was his main temple, where he regularly performed a Black Mass in black robes with a group of Crowley's followers. One visitor recalled how 'Two women in diaphanous gowns would dance around a pot of fire, surrounded by coffins topped with candles . . . All I could think at the time was if those robes caught on fire the whole house would go up like a tinderbox.' Sexual magic and drugs play an important role in Thelemic ritual magic, due to their ability to create changes in consciousness. Parsons wrote a poem called 'Oriflamme', which began, 'I hight Don Quixote, I live on peyote / marihuana, morphine and cocaine. / I never knew sadness but only a madness / that burns at the heart and the brain.'

Parsons's great occult project was to destroy the current world, which he viewed as patriarchal and corrupt, by unleashing an overpowering wave of dark female energy. To this end he embarked on a lengthy series of rituals aimed at manifesting the Biblical Whore of Babylon (or 'Babalon' as he preferred to name her). He was aided in this endeavour by the science fiction writer L. Ron Hubbard, who would later found the Scientology organisation. Their relationship did not end well, and Hubbard eventually abandoned Parsons. He took with him Parsons's lover and a large amount of his money, with which he bought a number of yachts. Parsons retaliated by declaring magical war and claimed to have summoned the sudden sea squall that almost sank Hubbard's boat. This messy situation led to a distinct cooling in Crowley's opinion of the pair. Referring to Parsons and Hubbard in a letter to his colleague Karl Germer, Crowley wrote that 'I get fairly frantic when I contemplate the idiocy of these goats.'

Parsons believed that, exactly seven years after his Babalon Workings, Babalon herself would manifest and rule over this world. His prophecy included a proviso that this would only occur if he was still alive. But during the Babalon Workings, Hubbard had channelled a dreadful warning, one that left him 'pale and sweaty': 'She [Babalon] is the flame of life, power of darkness, she destroys with a glance, she may take thy soul. She feeds upon the death of men. Beautiful – horrible . . . She shall absorb thee, and thou shalt become living flame before she incarnates.'

An explosion ripped apart Jack Parsons's home on 17 June 1952. It could be heard nearly two miles away. Parsons was at the very heart of it. His right arm was never found, so must have been closest to the source of the explosion. Much of his right jaw was gone, his shoes were shredded by the blast and his remaining limbs were shattered. He was found alive by neighbours, in the rubble of the firestorm. Papers both technical and occult floated through the air. He died thirty-seven minutes later, and was still just thirty-seven years old. His last words were, 'I wasn't done.' Fitting for an Antichrist perhaps – the opposite of Jesus's 'It is finished.'

Parsons, who had found work making explosives for Hollywood movies, had most likely made a careless mistake when working in his laboratory, where he stored a significant amount of chemicals and explosives. There have been a number of other theories for his death over the years, as you might expect given his links to military secrets and the world of the occult.

Parsons's interest in both rockets and ritual magic might seem surprising today, when a career in rocket science is considered to be professional and respectable. But when Parsons set out on his path, they were both fantastical. Such was Parsons's will and personality that he pursued them both regardless, in the long tradition of magically minded scientists such as Isaac Newton or the sixteenth-century astronomer John Dee. He was concerned with summoning and controlling colossal amounts of explosive energy, both mental and chemical. Both were dangerous, and in both he could be reckless. If he didn't have the demons that drove him towards the occult, it seems unlikely that he would have had such success in the field of rocket science. It was the same urge that propelled him down both paths.

In 1972 the International Astronomical Union named a crater on the moon 'Parsons Crater', to mark his pioneering work in the field of rocket science. It is perhaps apt that the crater that honours Marvel Whiteside Parsons is on the dark side of the moon.

In July 1943 a charming, handsome German aristocrat named Wernher von Braun was driven to Wolfsschanze, a secret bunker headquarters hidden deep in the woods outside Rastenberg in East Prussia. This was Hitler's infamous 'Wolf's Lair'. Von Braun's intention was to convince his Führer to support the production of a new rocket he had developed, the A-4.

Amidst the dark, grimy industry of the Second World War, A-4 rockets appeared to be technology from the future. Standing forty-six feet tall, these sleek, beautifully shaped rockets looked like illustrations from pulp science fiction. They had a range of two hundred miles and were decades ahead of anything the Russians or Americans were capable of building. The problem was that Hitler had previously dismissed the project because of a dream he once had. In that dream Hitler was convinced that no rocket would ever reach England, and for that reason he simply did not believe that the project was worthwhile.

By 1943 Hitler was pallid from living in bunkers away from sunlight. He looked much older and frailer than before the war, and the slight stoop as he walked made him look smaller. An army general who accompanied von Braun was shocked at the change in him. Yet as Hitler watched von Braun's confident presentation and saw footage of successful A-4 test flights, a marked change came over him. 'Why could I not believe in the success of your work? Europe and the rest of the world will be too small to contain a war with such weapons. Humanity will not be able to endure it!' he declared. Hitler took a gamble on von Braun's work winning him the war, and redirected much needed resources to the rocket programme. The A-4 would be renamed the V-2, the 'Vengeance Weapon'. Hitler wanted the rocket's load to be increased from a 1-tonne to a 10-tonne warhead, and he wanted thousands of them to be mass-produced every month.

For Hitler, the V-2 was a weapon that would break the spirit of British resistance in a way that his previous bombing blitzes had not. It probably would have done if it had been deployed earlier in the war, rather than during the European endgame when the Russians were approaching Berlin as the Allies advanced from the west. The V-2s that were used against England took a psychological toll on the war-weary population, which was noticeably greater than the impact of the 1940 and 1941 Blitz. There had been a ritual pattern to life under the Luftwaffe bombing campaigns against cities such as Coventry, Belfast and London. They began with the wail of the air-raid sirens, which heralded the increasing drone of the approaching bombers and the journey to the air-raid shelters until, finally, the all-clear was sounded. The British people had proved remarkably able to adjust to life under this ritual. The V-2s, in contrast, fell silently. They were undetectable in flight and impossible to shoot down. They could hit anywhere at any time, leaving the populations of target cities permanently unnerved and scared.

Bill Holman, a child survivor of the V-2 attacks, later remembered one of the 1,115 V-2 bombs which fell on or around London. 'On 24 November I was at junior school when all of a sudden it was rocked by a tremendous explosion. Rushing home, I met my friend Billy emerging from his front door, dazed yet calm, and announcing, "Mum and dad are dead." A clock tower had stood near our home. Now, hit by a V-2, it was replaced by a vast crater . . . Mr and Mrs Russell ran a vegetable barrow in the street; he was dead, she'd had her legs blown off. Mrs Popplewell, a friend of my mum's, was lifeless without a scratch on her. She had been walking when the blast from the rocket entered her lungs and she couldn't breathe. Over the road, a young soldier was on leave with his wife and mother. The wife had popped out to the shops and lived. Mother and son were killed.'

For von Braun, this was all incidental to the pursuit of his childhood dream. Like Jack Parsons, von Braun's goal was space flight. As a boy growing up in a wealthy Berlin family he had studied the moon through a telescope and strapped rockets to his go-kart. But space flight was as much a crank's dream in pre-war Germany as it was in pre-war America. The establishment did not take it seriously and it was certainly not something they funded. His only option was to keep that dream hidden while attempting to advance it through the only means possible: weapons research.

Von Braun did not show signs of being unduly troubled by the morality of this path. He joined the Nazi Party in 1937 and the SS in 1940. He was promoted every year until he reached the SS rank of *Sturmbannführer*. Following his successful appeal to Hitler, he oversaw production of his V-2 rocket at Mittelwerk, a factory built underground in order to protect it from Allied bombers. Mittelwerk consisted of 42 miles of tunnels hewn out of the rock by slave labour, and the descriptions of the state of the slaves when they were liberated by American forces in 1945 are harrowing. Over twenty thousand slaves died constructing the factory and the V-2. Mass slave hangings were common, and it was mandatory for all the workforce to witness them. Typically twelve workers would be arbitrarily selected and hung by their necks from a crane, their bodies left dangling for days. Starvation of slave workers was deliberate, and in the absence of drinking water they were expected to drink from puddles. Dysentery and gangrene were common causes of death. There are records of the liberating forces' failed attempts at removing the stench of death from the tunnels with strong disinfectant.

Von Braun himself was personally involved in acquiring slave labour from concentration camps such as Buchenwald. Ten years later he was in America presenting children's programmes on the Disney Channel, in an effort to increase public support for space research. Whatever he had to do in order to advance his dream, von Braun did it.

Von Braun's journey from the SS to the Disney Channel was fraught with difficulties. As one of the most gifted rocket scientists in history, he was regarded as something of a prize by Russian, British and American armed forces. But with the Third Reich collapsing around him in the final weeks of the European war, this did not mean that he or his team would not be killed by mistake, or that their work would survive the chaos of regime change. Hitler's infamous 'Nero Decree' of 19 March 1945 complicated matters further, for it demanded that 'anything . . . of value within Reich territory, which could in any way be used by the enemy immediately or within the foreseeable future for the prosecution of the war, will be destroyed.' Von Braun and his men were clearly of value and were placed under armed guard by the SD, the security service of the SS. Their SD guards were under orders to shoot all the rocket scientists the moment they were in danger of being captured by either of the oncoming armies. Fortunately, von Braun's colleague General Walter Dornberger was able to persuade the SD major that he and his men would be hanged as war criminals if they complied with this order, and that their only hope of surviving the war was to burn their black uniforms and disguise themselves as regular German troops.

Von Braun had already decided that he wanted to surrender to the Americans. His argument was that America was the only country not decimated by the war, and hence the only country financially able to support a space programme, but it is also clear that his aristocratic background would not have been well suited to life in the Soviet Union. America, in turn, wanted von Braun primarily because they didn't want anybody else to have him. Arrangements were quickly made to bring von Braun to America, along with his designs, his rockets and about a thousand other Germans (members of his team, along with their family members). An operation to whitewash the files of von Braun and other prominent Nazis in the team began. Jack Parsons's old mentor von Kármán was part of this process. It was known as Operation Paperclip after the paperclips which were used to attach fake biographies, listing false political affiliations and employment histories, to their files. Following Operation Paperclip, even Nazis who were guilty of war crimes were eligible for life in the US.

Von Braun settled into an army research facility in El Paso, Texas, together with his team and their families. His hopes of commencing research into a space programme, however, soon faltered. The US did not expect another international war to erupt and saw no reason to develop weapons for one. It would be over a decade before von Braun's talents were put to use. Yet the chain of events which led to this started with the end of the Pacific War. Man's conquest of space owes much to a destructive force on a par with that sought by Jack Parsons at his most insane. This appeared in the skies above Hiroshima and Nagasaki in August 1945.

The first nuclear weapon used against an enemy country was an A-Bomb called Little Boy. The United States Air Force dropped Little Boy on Hiroshima on 6 August 1945. It killed an estimated 66,000 people immediately, and 69,000 as a result of injuries and fallout. The second and last nuclear weapon ever used in warfare fell on Nagasaki three days later. It was called Fat Man, and it killed a total of 64,000 people. Both bombs were carried across the Pacific in the belly of a B-29 'Superfortress', an enormous four-engine bomber with a combat range of over 3,000 miles. The B-29 was developed when the Americans feared all of Europe was going to fall to the Third Reich, meaning that any air strikes against Germany would have to have been launched from Canada or the US. The B-29's development, with hindsight, can be seen as a significant moment in American history. It has come to symbolise the rejection of the United States' historic isolationist policies.

The invention of nuclear bombs, as well as nuclear energy, was another of the unforeseen implications of Einstein's General Theory of Relativity. Einstein had been playing around with the mathematics of his work when a beautifully simple equation popped out, as if from nowhere. This equation was $E=mc^2$. 'E' represented an amount of energy, but curiously the 'm' represented mass and so referred to physical matter, rather than energy. 'C' was a constant which stood for the speed of light. This was a big number, and once it was squared it became massive. The equation, therefore, said that a small amount of mass was equivalent to a really huge amount of energy. The question then became how to free and utilise that energy, and splitting up heavier, unstable elements such as plutonium or uranium seemed to be the way to go.

The decision to use nuclear weapons against Japan is still controversial. Some see it as a war crime and argue that, as President Eisenhower wrote in his memoir, 'Japan was already defeated and dropping the bomb was completely unnecessary.' Others point to the Japanese ancient warrior tradition of *bushido* as evidence that the country would never have surrendered, and to the fact that the bombing prevented a land invasion and hence saved the lives of many thousands of Allied troops. More recently historians have argued that the bomb was used as a show of strength towards the Soviet Union, rather than as a means to defeat Japan.

But presenting Stalin with a show of strength might not have been the smartest move on the board. To those with an understanding of Stalin's character it was clear that, after the Americans had demonstrated their power at Hiroshima, nothing on earth was going to stop him from obtaining a nuclear weapon of his own. He achieved this goal with impressive speed, relying as much on the skills of Russian intelligence agents, who stole Western nuclear secrets, as he did on the talents of Russian engineers. Russia detonated its first successful nuclear test in August 1949, and its first hydrogen bomb in 1953.

It is, perhaps, fortunate for the world that Russia caught up so quickly. A leading member of the US Atomic Energy Commission after the war was a Hungarian mathematician called John von Neumann. Von Neumann had been a child prodigy, able to divide two eight-digit numbers in his head at the age of eight and who simultaneously took degrees at three different universities at the age of eighteen. Von Neumann was a genius, and he became one of the

President's most respected advisers. His advice was this: Eisenhower had no choice but to immediately launch a massive unprovoked nuclear strike against Russia, and to nuke the Soviets back to the Stone Age before they developed a nuclear bomb of their own.

This wasn't just a whim, von Neumann explained. He had hard, cold, logical proof to back up his words. He had developed game theory, which we previously mentioned in Chapter 4. This is a field of mathematics, further developed by mathematicians such as John Nash, which models the actions of two self-interested parties. Game theory deals with situations where the moves of your opponents cannot be accurately predicted, where there is an absence of trust, and where damage limitation is a more rational goal than outright victory. Game theory fitted the Cold War stand-off perfectly, and its unarguable logic was that the only rational move was to immediately murder hundreds of millions of innocent Russian civilians. Certainly Eisenhower's secretary of state John Foster Dulles was convinced, and he pressured the President to send the bombs flying straight away. Eisenhower couldn't refute the logic, but he still felt that maybe this wasn't a good idea. He prevaricated long enough for Stalin to announce that he too had a hydrogen bomb, at which point von Neumann's logic collapsed.

Having nuclear bombs was one thing, but they were not weapons you would use locally. The immediate solution to this problem was to use giant flying fortresses such as the B-29 or its successor, the uncomfortably named B-36 Peacemaker, to drop those bombs on countries far away. But long-range transport planes are slow, noisy, and relatively easy to shoot down. What if rockets such as von Braun's V-2s could be fitted with nuclear warheads? If those rockets could become reliable enough to travel thousands of miles and still accurately hit targets such as Moscow, Beijing or New York, then those cities would be as defenceless as London had been in the dying days of the Second World War.

For America in particular, this was psychologically very difficult. The United States had never really regarded itself as being at risk before. It had fought bravely in two world wars, but the only damage it had suffered on the home front was the Japanese bombing of the navy base at Pearl Harbor. This occurred on the Hawaiian island of Oahu, nearly four thousand miles from the North American mainland, so it impacted on the typical American in a different manner to the bombardment and destruction that many European populations had come to terms with. But suddenly the United States could be destroyed, with weapons it had itself developed, by a powerful dictator on the other side of the world. The arrival of the mushroom cloud rewrote the existing geopolitical game. The United States was a free, functioning democracy and yet the President could still be advised by someone as crazy as von Neumann. No one wanted to imagine what sort of advice a sociopath such as Stalin might be receiving.

As fate would have it, key engineers in both the American and Russian rocket programmes still privately regarded their weapons work as a front for their boyhood dream of space travel. Persuading their respective governments to commit the tremendous amount of cash needed to achieve this dream was unrealistic in the immediate postwar era. But building a rocket of sufficient power, reliability and accuracy to nuke the other side of the globe was, as it turned out, an identical engineering problem to that of building a rocket that could leave earth's gravity and enter space.

Vast expenditure on weapons technology didn't generate the sort of public image that leaders wanted to project. Declaring your futuristic-sounding achievements in peaceful space flight was much more attractive. Having such a rocket was one thing, but it would only work as a deterrent if the other side knew that you had it. Announcing the development of the technology necessary to enter space was a coded way of saying that you were able to nuke every corner of the planet.

They came for Sergei Korolev at 9 p.m. on 27 June 1938. He was with his three-year-old daughter in their sixth-floor Moscow apartment when his wife rushed in, panicking. She had been downstairs and seen a number of NKVD officers entering the building. She knew instinctively that they were coming for her husband. Korolev worked as a rocket scientist, and military research was as riddled with fear and paranoia as the rest of the Soviet system. His friend and colleague Valentin Glushko had recently been taken. It had to be assumed that he would have denounced anyone his torturers suggested. Such was the way of life under Stalin's Great Purge.

The NKVD promptly arrived, ransacked the apartment, and took Korolev from his family. Two days later, after torture and threats against his family, he signed a confession to the Commissar of Internal Affairs admitting his involvement in a counter-revolutionary organisation and to committing acts of sabotage against the Motherland. He confirmed allegations against him made by two of his senior colleagues, who had since been shot. He had no trial, but was given a ten-year sentence in the notorious Kolyma Gulag in Siberia, at the fringes of the Arctic Circle.

It probably doesn't need saying, but there was no counter-revolutionary organisation, no act of sabotage, and Korolev was entirely innocent. Many people were innocent during the Great Purge. Reason and justice played no part in that terror.

It was a miracle he survived. Thousands died every month at Kolyma. Korolev was starved and beaten, his teeth were knocked out and his jaw broken. The bitter cold was unbearable, and the malnutrition led to scurvy. He was, ultimately, rescued by far-off events. The appointment of a new head of the NKVD, Lavrenti Beria, had resulted in a number of his predecessor's cases being re-opened. Korolev was told to leave the gulag and report to Moscow, where he would be retried on lesser charges.

There was no transport available, so he was forced to hitch his way home. A lorry driver took him a hundred miles to the port town of Magadan on the Sea of Okhotsk, but they arrived too late to catch the last boat of the year. In payment, the lorry driver took his coat. Korolev was left to survive as best he could, malnourished and with thin clothing that froze to the floor when he slept.

Korolev remained in Magadan for the winter, while temperatures fell as low as minus fifty degrees Celsius. In the spring he was able to sail to the mainland and catch a train to Moscow, but he was removed from the train at Khabarovsk for being too ill to travel. Korolev would have died shortly afterwards had he not been found by an old man who took pity on him and nursed him better. A few weeks later he was lying under a tree feeling the warmth of spring on his skin. Korolev opened his eyes and saw an exquisitely beautiful butterfly fluttering in front of him. That was the moment when he realised he was going to live.

On 13 May 1946 Stalin, reacting to the American use of the atomic bomb in Japan, issued a decree that ordered the creation of a scientific research institution called NII-88. This was a rebirth for Soviet rocket research, which had ground to a halt following the purges of the 1930s. Korolev found himself once again working in the field that, like Parsons and von Neumann, he had been driven to by childhood dreams of space flight.

The first order of business was for the Russians to learn what they could from the Nazis. Sergei Korolev arrived in Berlin one day after Wernher von Braun left Germany for good. The Americans had stripped the Nazi rocket factories at Mittelwerk and Peenemünde of anything that they thought would benefit the Russians, but they did not reckon on the tenacity of the Soviets or the brilliance of Korolev. The workings of the V-2 were painstakingly pieced together from scraps and clues. Korolev was able to see how the V-2 worked. More importantly, he was also able to see its faults. Korolev looked at von Braun's great achievement, and thought that he could improve it.

During the 1950s distrust between East and West became institutionalised. Anti-communist paranoia in America led to the inquests of Republican Senator Joe McCarthy. The McCarthy 'witchhunt', which led to smears, blacklists and book-burning, greatly troubled Einstein. He had lived through the rise of the Nazis and it looked to him as if history was starting to repeat itself. Liberal democracy has the ability to correct its excesses in a manner that did not happen during the Weimar Republic, but Einstein did not live to see this. He died in Princeton hospital in 1955 at the age of seventy-six, much loved and admired, but troubled by the complicated global politics his theories had in part created.

During these years von Braun became increasingly frustrated by the way the United States Army kept him and his team treading water. They had no interest in funding the expensive development needed for von Braun to achieve his dreams of leaving the planet. His collaboration with Walt Disney in the mid-1950s came about due to his desire to increase interest in space research, in the hope that public support would translate into research money. His TV programmes made him the public face of space travel, but there was no love for this ex-Nazi in army or government. Research projects that he would have been ideally suited for were given to the Navy or the Air Force, or to other American engineers in the army.

Korolev, meanwhile, worked hard. He regained the trust of Stalin, and convinced him of the importance of long-range rocket technology. By 1947 he had built and tested copies of the V-2. In 1948 he began work on a rocket of his own design, the R-2, which was more accurate and had twice the range. He successfully tested this in 1950. In 1953 he convinced the Kremlin to back his plans for an even larger rocket, the R-7, which would be powerful enough to launch a satellite into orbit. All this was achieved while he was under extreme pressure for results, in a country devastated by war, and in an economy organised under Stalin's unworkable state-controlled 'Five-year Plans'. Korolev's time in the Gulag had taught him to survive, but he was working himself to death.

Both von Braun and Korolev began framing the conquest of space in a way that spoke to the paranoid military mind. The nation that first controlled space would be all-powerful, they argued. Permanent orbiting space stations could spy on every aspect of the enemy, and even drop bombs on any point of the globe. If we didn't develop this technology, they said, then surely the other side would. Korolev convinced Stalin and his successor Khrushchev that rocket development had reached the point where all this was feasible, but von Braun was still viewed with some suspicion by the US government, despite his growing public fame.

America officially entered the space race in 1955, when Eisenhower announced plans to mark 1957's International Geophysics year by launching a satellite into orbit like 'a second moon'. But Eisenhower wanted Americans to achieve this landmark, not Germans. To von Braun's frustration, the task of launching the satellite was given to the United States Navy.

Korolev also planned to launch a satellite in 1957. His satellite was a metal ball 23 inches in diameter, which had four elongated trailing radio aerials. It was made from a polished metal that made this simple shape look exciting and futuristic. It is, strangely, a remarkably beautiful object. Korolev recognised this, and insisted that it had to be displayed on velvet. Its name was Sputnik 1.

Sputnik was technically very simple. Apart from a few systems to monitor temperature and pressure, it consisted of little more than a radio transmitter that made regular pulses on two separate frequencies. The original plan of launching a more sophisticated machine was scrapped due to overwhelming pressure from the Kremlin to get something in orbit before the United States. A more elaborate satellite had been designed and built, but it could not be made to function reliably in time. It did not help that the driver who transported that satellite across Russia drank a substantial amount of industrial alcohol and crashed the truck into a tree.

On 4 October 1957 Sputnik 1, a metal ball that did little but go bip-bip-bip and the last-minute replacement for a proper satellite, became the first object the human race launched into orbit. All over America, indeed all over the world, people could tune their short-wave receivers into the correct frequency and hear the pulses it broadcast as it flew overhead. Most of the world congratulated the Soviet Union on its historic achievement, but for the American public this was a terrifying, unexpected shock. The Russians were above them.

The gloves were off. Suddenly there was no shortage of political will to back the space programme. Unfortunately for the Americans Korolev had no intention of slowing down, despite signs of heart problems. Sputnik 2 launched less than a month after Sputnik 1. It contained a dog called Laika, who became the first living creature launched into space. The satellite's cooling system failed and Laika was cooked to death during the flight, but this was not announced at the time and Laika became a folk hero to the Soviet people. The first spacecraft to land on another celestial body was Luna 2, which reached the moon on 14 September 1959. It is probably fairer to say that it crashed into the moon rather than landed on it, but that was sufficient to achieve its aim of delivering an engraving of the Soviet coat of arms to the moon's surface. Khrushchev gave Eisenhower a copy of this pennant as a deliberately tactless gift. A further Luna probe in October 1959 sent back the first pictures of the dark side of the moon. On 19 August 1960, two dogs named Belka and Strelka became the first living things to journey into space and, crucially, return alive. Then, on 12 April 1961, a farm boy from war-ravaged western Russia, the son of a carpenter and a milkmaid, became the first person in space. His name was Yuri Gagarin.

The Soviet Union had won the space race.

'I am watching the earth . . . I am feeling well,' he reported from orbit. 'The feeling of weightlessness is interesting. Everything is floating! Beautiful. Interesting . . . I can see the earth's horizon. It is such a pretty halo . . . I am watching the earth, flying over the sea . . .'

Gagarin was unflappable and good-natured with a handsome, boyish face. He was the ideal poster boy for a country still traumatised by the losses it suffered during the German invasion, and he helped restore Russian confidence and pride. He became a global celebrity overnight.

Before the flight, Korolev and his team realised that they needed a word to describe the capsule, Vostok 1, which Gagarin would pilot. After much debate they decided to use the word 'spaceship'. All this occurred before the development of modern digital computers and only twenty-two years after the first flight of a jet-engine-powered plane.

The American space programme was not going well. The Navy's attempts to launch a satellite in 1957 failed, so von Braun suddenly found himself in favour. Wasting no time, he was able to place the first American satellite in orbit in January 1958, but this was just one small success in a humiliating string of disasters. The rush to catch up meant that many American rockets either failed to launch or exploded spectacularly. It did not help that every American failure occurred in the full glare of the public gaze, while Russian failures were routinely hushed up. The newspapers were quick to headline such failures with terms like 'Kaputnik!' A rocket that only succeeded in reaching four inches off the ground was a particular embarrassment. When news of Gagarin's flight broke, an American reporter rang the newly established American space agency NASA for their reaction. It was 5:30 a.m., and the phone was answered by a PR man who had worked long into the night and slept in his office. 'Go away,' he told the reporter. 'We are all asleep.' The reporter's headline became: 'SOVIETS PUT MAN IN SPACE. SPOKESMAN SAYS U.S. ASLEEP.' When America did succeed in placing their first man in space, on 5 May 1961, it was notable that they could only achieve suborbital flight. Unlike Yuri Gagarin, who flew around the entire planet, the American astronaut Alan Shepard went straight up and, fifteen minutes later, came straight back down again.

In December 1960 Sergei Korolev suffered his first heart attack. He recovered and returned to work, where he continued to push himself far beyond the boundaries that his health could stand. His health deteriorated, and internal bleeding and intestinal problems joined his cardiac arrhythmia. He had faced death before, during his time in the Gulag, and he refused to let it interfere with his work. He was able to notch up further historic achievements, such as launching the first woman into space in June 1963 and the first spacewalk, in March 1965. But in January 1966 he died, and the Russian space programme collapsed around him. The era of Russian space 'firsts' came to an abrupt halt.

Korolev was unknown during his lifetime. His identity was kept secret, for fear of American assassination, and he was known to the public only by the anonymous title 'Chief Designer'. With death, however, came fame. His ashes were interred with state honours in the Kremlin Wall, and history now recognises him as the architect of mankind's first steps into the cosmos. In time, if mankind does have a future out among the stars, he may well come to be remembered as one of the most important people of the twentieth century.

On 25 May 1961, barely six weeks after Yuri Gagarin became the first man in space, the new American president, John F. Kennedy, addressed Congress: 'I believe that this nation should commit itself to achieving the goal, before the decade is out, of landing a man on the moon and returning him safely to Earth,' he said. 'No single space project in this period will be more impressive to mankind, or more important for the long-range exploration of space, and none will be so difficult or expensive to achieve.'

This project was the Apollo programme. It was, in many ways, a crazy idea. The space race had already been lost. Kennedy was correct in recognising how 'difficult and expensive' the project would be, but he should also have mentioned 'dangerous'. Three crew members of Apollo 1 would burn to death in a cabin fire in January 1967.

If the aim of the project had been to collect moon samples for scientific study, then the Russian approach was more sensible. They sent an unmanned probe called Luna 16 to the moon, which drilled down, collected rocks, and returned them to earth. If the aim of the project was to drive around the surface in a little buggy taking pictures, then the Russians did this too, with the unmanned Luna 17. The American manned approach was so dangerous and expensive that it seemed impossible to justify.

But the aim of the Apollo programme was not simply scientific. It was an expression of the American character, which would come to typify the second half of the twentieth century. So what if the space race had been lost? If the public could think of that race in radically different terms, then it could still be won. The goal posts could be moved.

The United States understood what people wanted. They wanted to see men climb into a spacecraft, land it on the moon, then walk out and hit golf balls around. They wanted to see astronauts drive space buggies across alien landscapes, then send messages of love home to their wives and children. The hard, useful science that Korolev had achieved was all well and good, but what people really wanted was what the B-movies and science fiction comics had promised. The Apollo programme, if it could succeed, would rewrite the nature of the space race in people's minds. It would redefine it not as something that had been lost, but as something that they were always destined to win. This was extraordinarily risky because what Kennedy asked NASA to do was so difficult that it did not, at the time, appear possible. It was the most expensive 'Hail Mary pass' in history.

As Kennedy's speech made clear, his reasons were political rather than scientific. 'If we are to win the battle that is now going on around the world between freedom and tyranny,' he told Congress, 'the dramatic achievements in space which occurred in recent weeks should have made clear to us all, as did the Sputnik in 1957, the impact of this adventure on the minds of men everywhere, who are attempting to make a determination of which road they should take.' The heroic flight of Yuri Gagarin, in other words, might lead people towards communism. If Western democracy was a superior system to the unworkable horror of communism, then what were the Russians doing orbiting the planet and racking up achievements that were, frankly, beyond the ability of American engineers?

The solution to the problem was to throw money at it, and the aim of Kennedy's address was to secure massive amounts of taxpayers' dollars. 'Our greatest asset in this struggle is the American people,' Kennedy said, 'their willingness to pay the price for these programs, to understand and accept a long struggle, to share their resources with other less fortunate people, to meet the tax levels and close the tax loopholes I have requested.' It is hard to imagine a President using those words today. Kennedy asked for $7 billion, but the final cost was over $25 billion. It is one of the many ironies of the American space programme that, even though it was intended

to demonstrate that the American system of freedom and individualism was superior to communism, it could better the achievements
of a single-minded Russian genius only through an expensively
funded government programme.

This was von Braun's hour. While Korolev had toiled away in obscurity, von Braun became the face of the US space programme.
His profile had already been boosted by a 1960 film of his life story,
somewhat sanitised, entitled *I Aim at the Stars*. The film was re-
titled *Wernher von Braun* for its British release, possibly to avoid
the common joke that the film's full title was 'I Aim at the Stars
(but Sometimes I Hit London)'. Yet regardless of his past, there was
no doubt that von Braun was the man for the job. The towering
Saturn V rocket he and his team created was a modern wonder of
the world. It rose elegantly into the Florida sky on 16 June 1969, with
the comparatively minuscule Apollo 11 spacecraft in its nose. Inside
that craft sat the best of the best, the astronauts Michael Collins,
Buzz Aldrin and Neil Armstrong.

Mission control began a 'T-minus' countdown, with launch occurring when the countdown reached zero. This idea was taken
from Fritz Lang's 1929 film *Frau im Mond* (Women on the Moon).
Von Braun was a big fan of this film, and had painted its logo on the
base of the first V-2 rocket launched from Peenemünde.

Four days later, after a journey of 384,400 kilometres, Armstrong
became the first human being to set foot on a celestial body other
than the earth. It was one small step for a man but, as he so perfectly
summed up the moment, one giant leap for mankind.

It was also a giant leap for America. In the nineteenth century, it
had been the British who led the way in culture, science and progress. In the first half of the twentieth century, Germany and the
German-speaking European countries had taken that role. After
those nations had been stricken low by the cancer of fascism, two
giant superpowers emerged as contenders for the title of the world's
leading nation. In the nuclear-war-poised geopolitics of the Cold
War, the only safe arena for them to compete was off-world. When
Armstrong's boot crunched into the fine grey dust of the lunar

surface, the world had a winner. The twentieth century became, and will always be known as, the American Century.

The Apollo programme worked on a level above military, political or scientific advancement. It was not just Jack Parsons, Sergei Korolev and Wernher von Braun who grew up reading pulp science fiction and dreaming of making it a reality. The singular determination that those men demonstrated may have been rare, but their dream was shared by countless others.

On 14 December 1972, the crew of Apollo 17 left the moon. They did not realise it at the time, but they would be the last people to travel outside the earth's orbit for at least half a century. There is hope that the Chinese or private companies may visit the moon or even Mars in the twenty-first century, but it is also possible that mankind will never return. Once the political aims of the Apollo project had been achieved, the argument for government funding of space research on that scale collapsed. Ambitions dropped back down to the level displayed by the Russians in the 1950s and 1960s, when unmanned machines performing valuable science was financially justifiable, while the human dream of exploration for its own sake was not. This could be seen as a massive anticlimax, except for one thing.

In December 1968 the crew of Apollo 8 became the first humans to leave earth's orbit and travel into space. They launched with the intention of orbiting the moon and becoming the first people to look upon its dark side with their own eyes. They achieved this, but they also saw something else. It was not something that they had been expecting, but it turned out to be something of the utmost importance. When they rounded the dark side of the moon, the crew of Apollo 8 became the first humans to see the whole of planet Earth, hanging alone in space, blue and white and indescribably beautiful. They photographed it, and called that photograph *Earthrise*.

In 1948 the English astronomer Fred Hoyle, who coined the term the Big Bang, predicted that 'Once a photograph of the Earth taken from outside is available, a new idea as powerful as any in history will be let loose.'

The earth that Apollo 8 left was massive and entirely dependable, yet there it was, infinitely small and shockingly delicate. The sixty-two miles of atmosphere, which had looked an insurmountably massive barrier to engineers like Parsons, Korolev and von Braun, were seen as a fine, delicate wisp hugging the surface, a simple line separating the wet ball of rock from the void.

In the twentieth century mankind went to the moon and in doing so they discovered the earth.

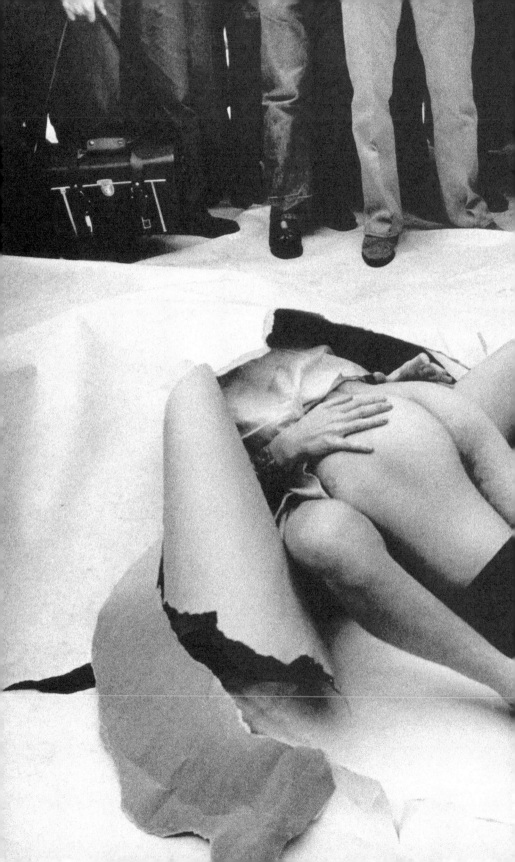

Live performance art in Soho, New York, 1970 (Jill Freedman/Getty)

Nineteen sixty-three
(which was rather late for me)

In turbulent periods of history a person can go from being a conservative, to a dangerous radical, to an embarrassing reactionary,
without once changing their ideas. This is what happened to the
English palaeobotanist Dr Marie Stopes in the early decades of the
twentieth century.

Marie's mother, the Shakespeare scholar Charlotte Carmichael
Stopes, was the first woman in Scotland to take a university certif-
icate. This was at a time when women were not allowed to attend
lectures or receive degrees. Charlotte wrote many academic works
on Shakespeare but her most successful book was *British Freewomen:
Their Historical Privilege* (1894), which was an inspiration to the
twentieth-century suffragette movement. Charlotte was a strong
feminist who campaigned for women's suffrage.

Unlike her mother, Marie Stopes didn't initially appear to be ac-
ademically inclined. Her formal education only began at the age of
twelve, when she was sent away to a suffragist-founded Edinburgh
boarding school. But despite her late start she applied herself and
eventually enrolled at the Botanical Institute at Munich University.
She was, at first, the only woman among five thousand men. From
here, at the age of twenty-four, she surpassed her mother's academic achievements and attained her doctorate. Dr Stopes became
a recognised expert in plant fossils, and seemed on course to a life
dedicated to the study of coal.

Like her mother, Marie was a great believer in female education
and political equality. But while Charlotte was in favour of the radical activism promoted by Emmeline Pankhurst, Marie preferred
more conservative suffragists. Pankhurst was exasperated by the

failure of the women's suffrage movement to produce tangible results in the nineteenth century, so she advocated the confrontational approach now known as direct action. Her followers shouted down politicians, chained themselves to railings, and committed arson. They threw stones through the windows of Buckingham Palace which were attached to notes explaining that 'Constitutional methods being ignored drive us to window smashing.' The death of Emily Davison, who was trampled to death at the Epsom Derby after stepping in front of a horse owned by King George V, became the defining image of Pankhurst's direct-action movement.

For Marie Stopes, this was all a bit much. Both she and her mother agreed that women needed the vote. They understood that this would lead to equality in a whole range of areas, from female-instigated divorce to tax status. Their difference of opinion was down to temperament. Marie did not think that direct action was in any way ladylike. She was not initially persuaded by her mother's argument that a lady who found a burglar in her house would be quite correct in hitting them over the head with a broom, so this was also correct behaviour for women who had been robbed of their political rights.

Yet there was one area where Marie was more radical than Charlotte, and that was sexually. Charlotte was the product of a society where wives were obliged to submit to their husbands for procreation, but were on no account to admit any sexual feelings themselves. After providing her husband Henry with two children, as was deemed proper, Charlotte believed she had fulfilled her wifely duties and withdrew physically from her husband. Henry Stopes found this lack of intimacy difficult. In an 1886 letter he wrote of his hope that they would next meet 'with the scales taken from your eyes as to the effects and need of greater love existing between us . . . Dearest, will you put from you the teachings of your splendid brain and look only into the depths of your heart and see if you can but find there the love that every woman should hold for the father of her babes? We would put from us the seven blank years that are ended and commence the truer honeymoon.' But from the formal tone of her letters to him, it appears that such pleas had little effect. Henry's letters after this point gradually give up hope of a more emotional and physical marriage. He died an early death a few years later.

Marie, in contrast, was more passionate. The poetry she wrote throughout her life constantly skirted around the borders of erotica. She saw physical union in spiritual terms, and viewed love between heaven-matched equals as the pinnacle of Christian society. This perspective was purely theoretical, at least in the first half of her life. She claimed to be entirely ignorant of the existence of homosexuality and masturbation until the age of twenty-nine, and she was still a virgin at the age of thirty-eight when she wrote her most famous book, the million-selling *Married Love*, which was eventually published in 1918.

Married Love was the result of the failure of her unconsummated first marriage in 1911, which only lasted two years. Her husband Reginald moved to Canada after the relationship failed, and Marie instigated divorce herself. Finding herself in the position of having to properly construct her legal case, and not knowing how unusual an unconsummated marriage was, she headed to the library to research human sexuality. Science and academia, she soon realised, had very little to say about the sex lives of husbands and wives. As someone who herself had written about reproduction in extinct plant life, she found this surprising. It struck her that writing about human sexuality might be socially useful.

In common with many in her class, and even with the benefit of her education, Stopes had embarked on her first marriage blind to the realities of sexuality. Such ignorance was then seen in terms of the desired quality of 'innocence', and hence widely promoted. But ignorance, she discovered, was not bliss, and her marriage failed to live up to her romantic dreams of true union. Not wishing others to go through the same experience, she began to write a manual for newlyweds based in part on what she learnt in biology textbooks, and in part on what she felt in her romantic heart.

Stopes argued that sexual fulfilment was necessary for the

physical, spiritual and emotional wellbeing of women. It was of vital importance, therefore, that husbands learnt to properly seduce their wives, and understand their monthly cycle of arousal. The responsibility fell to men, Stopes explained, because it was in no way proper for women to instigate sex, or drop their husbands hints.

Married Love was followed by an even more controversial book, *Wise Parenthood: A Book for Married People* (1918), which dealt with contraception. While sexual intimacy between spouses had value in itself, it could lead to women having a child a year for much of their adult life, whether they wanted to or not. For many women, this amounted to a form of physical and emotional slavery. The answer, as now seems obvious, was birth control. But this was at the time a highly contentious position. While *Married Love* had been attacked on the grounds that unmarried people might read it and be corrupted, *Wise Parenthood* faced a far greater range of enemies.

Promoting contraception wasn't officially illegal in Britain, but it risked prosecution under obscenity laws. At the time, Holland was the only country in the world where birth control was approved by the state. The first Dutch birth control clinic had opened in 1882. In contrast, the American nurse Margaret Sanger opened the first American birth control clinic in New York in 1916 and was promptly arrested. She was charged with distributing contraceptives and 'running a public nuisance'.

The average UK family had 2.8 children in 1911, compared to 1.7 in 2011. Large families were considered admirable. In the twenty-first century, tabloid newspapers routinely condemn large working-class families, but in 1921 the *Daily Express* ran a competition to find Britain's biggest family. It offered a £25 prize for the winner. Politicians, doctors and members of the public – particularly women – spoke out loudly against birth control. Doctors made a good deal of money from attending pregnant women, which the widespread adoption of contraception threatened. It was socially safer to adhere to the established cultural position, namely that sex was allowable for the purposes of procreation within marriage but was otherwise obscene. A person departing from that script risked the implication that they themselves found pleasure in sexual intimacy, and hence were some form of deviant.

The wave of opposition that faced Marie Stopes and other birth control pioneers reminds us how much our understanding of our place in the world changed in the early twentieth century. In an age of individualism the idea that women have the right to make decisions about their own bodies appears self-evident, but in the imperial world it was the social hierarchy which dictated what should happen. And as the state religion is a mirror of existing social structures, it should be no surprise that the greatest opposition to Stopes's work came from the Christian Church.

Contraception was condemned by most Christian sects. In 1920, the Lambeth Conference of Anglican bishops called for the removal of 'such incentives to vice as indecent literature, suggestive plays and films, the open or secret sale of contraceptives, and the continued existence of brothels'. The strongest condemnation came from Catholicism. P.J. Hayes, the Roman Catholic Archbishop of New York, claimed in 1921 that contraceptives were worse than abortion. 'To take life after its inception is a horrible crime,' he argued, 'but to prevent human life that the Creator is about to bring into being is Satanic.'

The controversy came to a head when Stopes sued Halliday Sutherland, a Catholic doctor, for defamation, following criticism in Sutherland's 1922 book *Birth Control: A Statement of Christian Doctrine against the Neo-Malthusians*. Sutherland's defence was financially supported by the Church, so the case was widely seen as a fight between one woman and the Catholic Church.

Here Stopes's conservative nature came into play. Her appearances in court showed that she was clearly not the major threat to public decency that her opponents made her out to be. She was opposed to abortion and sex outside of marriage. She was charming and graceful, and her mode of dress was discussed in the press at length, to much approval. She avoided criticism that she was unladylike or a radical, so there was nothing to distract from her argument that women had the right to control their own reproduction. When she later gave a talk organised by a railway worker in Liverpool's Philharmonic Hall, that worker afterwards wrote to say that 'I would not like to meet you too often or I should fall in love with you – even if you are a Tory – because I admired your voice, your pluck and the way you handled your audience . . . Permit me too to compliment you upon your eloquence and the timbre, I mean sweetness, of your voice; it carries with it all that the word "feminine" ought to mean.'

The press interest in the case achieved what she had previously failed to do. It brought her crusade for birth control to the attention of the working class. Her books had sold well, but it was only the middle classes and above who could afford to buy books. When Stopes opened her first birth control clinic in 1921, it was significantly situated in Holloway, a working-class area of North London, and she wrote pamphlets specifically aimed at the poor as they, she felt, were the section of society who most needed to adopt contraception. Part of Stopes's argument was that middle-class doctors, clergy and journalists who opposed her were hypocrites: birth rates for the middle class were lower than for the working class, indicating that they had knowledge of birth control which they were denying to the poorer sections of society. It was thanks to press interest in her trial that Stopes's name and mission became widespread among the poorer section of society. It was even immortalised in a playground rhyme, 'Jeanie, Jeanie, full of hopes / Read a book by Marie Stopes / Now, to judge by her condition / She must have read the wrong edition.'

The result of the trial was messy. The judge awarded the case to Sutherland, in seeming opposition to the wishes of the jury. This was overturned by the Court of Appeal, which was in turn overruled following an appeal to the House of Lords. Three of five Law Lords involved in the case were over eighty years of age and their decision was 'scandalous', in the opinion of George Bernard Shaw. Nevertheless, in the wider world of public opinion, it was clear that Stopes's argument had triumphed. Birth control was discovered, and accepted, by the general population.

The trial would be the high point of Stopes's fame. Afterwards, she became increasingly vocal about her more reactionary beliefs. Birth control was a racial matter, she insisted. Inferior types were breeding faster than their betters, which was a situation that had to be reversed for the long-term survival of the white race. Mixed-race children should be sterilised at birth, as should all mothers unfit for parenthood. On the subject of eugenics Marie Stopes was as far right as Hitler, who began a programme of compulsory sterilisation of 'undesirables' in 1934. The forcefulness of her personality was accompanied by both an inability to admit mistakes and a need for praise. She made enemies easily, and it soon became impossible for her to work within the growing birth control movement.

Stopes's personality had touches of later radicals such as the psychedelic evangelist Timothy Leary or the computer-hacker turned whistleblower Julian Assange. All three managed to place previously unthinkable ideas right in the heart of public debate. They were all, briefly, lionised for their efforts. Yet the single-minded, messianic nature of their personalities turned the public from them and made their names toxic. Others would gain applause for work in the territory that they staked out, but it took a rare psychological character to introduce the world to that new territory in the first place.

For all that the name Marie Stopes has become tarnished over the years, she brought the concept of birth control to a wider public than anyone had previously managed. The value of sex, without intent of procreation, was finally admitted.

This was one factor in a larger revolution. Individualism required women to redefine both their sense of themselves and their position in society. The possible roles for middle-class women, beyond the traditional wife, mother and housekeeper, had been extremely limited in the patriarchal imperial world. In her 1929 essay *A Room of One's Own*, the modernist English writer Virginia Woolf recalled that 'I had earned a few pounds by addressing envelopes, reading to old ladies, making artificial flowers, teaching the alphabet to small children in a kindergarten. Such were the chief occupations that were open to women before 1918.'

Woolf's essay examined the topic of women and fiction, and

questioned why history had produced no female writers on a par with Shakespeare. She asked what would have happened to an imaginary sister of Shakespeare with the same innate talent as her brother. Shakespeare's sister, she concluded, would have been stifled at every turn by women's historical lack of financial independence and privacy, and the fixed expectations of hierarchical society. Female genius, Woolf believed, could not emerge until women were in a position to have a room of their own, where they could lock the door and remain undisturbed, and a personal income of £500 a year (just over £27,000 in 2015).

As both men and women gained freedom from what was expected of them at birth, the competition for rewarding and worthwhile careers increased. For men, that competition was lessened when women were encouraged to remain in their previous positions. Woolf noted this unwillingness to accept women as equals in historically male professions. 'The suffrage campaign was no doubt to blame,' she wrote. 'It must have roused in men an extraordinary desire for self assertion . . . When one is challenged, even by a few women in black bonnets, one retaliates, if one has never been challenged before, rather excessively.' For Woolf, the belief that the female half of the world was inferior gave men a valuable boost of self-confidence, and allowed them to go forth and achieve great things.

The pressure for women to remain in their historic roles became increasingly problematic as individualism grew. The American writer and activist Betty Friedan, who felt forced out of her career as a journalist when marriage and children turned her into a homemaker, wrote about the gnawing sense of undefined dissatisfaction that marked the lives of many housewives in post-Second World War America. This was the subject of her book *The Feminine Mystique* (1963), which examined how women could find personal fulfilment outside of traditional roles. 'American housewives have not had their brains shot away, nor are they schizophrenic in the clinical sense. But if . . . the fundamental human drive is not the urge for pleasure or the satisfaction of biological needs, but the need to grow and to realise one's full potential, their comfortable, empty, purposeless days are indeed cause for a nameless terror,' she wrote. As individuals, women needed a sense of purpose that was based around themselves, and not their husband or family.

For Friedan, feminism was about 'freeing both women and men from the burdens of their roles'. She went on to co-found the National Organisation for Women, and the success of her book triggered a new wave of feminist thought and activism. The acceptance of female sexuality, in post-Friedan feminist thought, was just one section of a broader conversation about the role of women in the age of individuals.

The English poet Philip Larkin dated the arrival of an acceptance of sex in British culture to a very specific point. Prior to this moment, he wrote in his poem 'Annus Mirabilis', sex existed only as 'A shame that started at sixteen / And spread to everything'. The moment when everything changed was 'Between the end of the Chatterley ban / And The Beatles' first LP'. This was the early Sixties, a time when press coverage of the Profumo scandal, when a government minister was revealed to have the same mistress as a Soviet naval attaché, reflected an increased sexual openness in the British public. Larkin, then in his forties, wrote that 'Sexual intercourse began / In nineteen sixty-three / (which was rather late for me).'

D.H. Lawrence was not a writer who was accepted or admired by critics in his own lifetime. His novel *Lady Chatterley's Lover*, which was written in 1928, a couple of years before his death, had to wait decades before it was openly praised. It was initially only published privately, or in heavily abridged versions, due to its sexual explicitness and taboo language. To many people, these seemed all the more shocking because they were delivered in the broad rural Nottinghamshire dialect of Mellors, the Chatterleys' gamekeeper. 'Let me be,' he says to Lady Chatterley after sex, 'I like thee. I luv thee when tha lies theer. A woman's a lovely thing when 'er's deep ter fuck, and cunt's good. Ah luv thee, thy legs, an' th' shape on thee, an' th' womanness on thee. Ah luv th' womanness on thee. Ah luv thee wi' my bas an' wi' my heart. But dunna ax me nowt. Dunna ma'e me say nowt.'

Later attempts to publish the complete text led to obscenity trials in countries including India, Canada and Japan. In the United States, the Mormon Senator Reed Smoot threatened to read passages from it aloud in the Senate. *Lady Chatterley's Lover* was, he declared in 1930, 'Most damnable! It is written by a man with a diseased mind and a soul so black that he would obscure even the darkness of hell!'

The 1960 British prosecution under the Obscene Publications Act that Larkin referenced in his poem followed an attempt by Penguin Books to publish the unabridged text. During the trial the chief prosecutor Mervyn Griffith-Jones famously asked the jury, 'Is it a book that you would even wish your wife or your servants to read?' This recalled the questioning during Marie Stopes's 1923 libel trial, where the expert medical witness Sir James Barr was asked if he thought the book could be 'read by your young servants, or, indeed, [would you] give it to your own female relatives?' This comment was unremarkable at the time, but the country had changed between 1923 and 1960 and Griffith-Jones's question came to symbolise how out of touch the British establishment had now become. Penguin were acquitted of obscenity and the publishing industry has had the freedom to print explicit material ever since.

The fact that Griffith-Jones's out-of-touch remark about servants came to represent the trial is, in many ways, entirely fitting for *Lady Chatterley's Lover*. The novel tells the story of Constance Chatterley, the young bride of the aristocratic Lord Clifford Chatterley. Lord Chatterley was seriously wounded in the First World War and returned impotent and paralysed from the waist down. He was the last of his line. His inability to produce an heir and continue his dynasty weighed heavily on him, because he understood the world through the pre-First World War hierarchical model. As he tells his wife, 'I believe there is a gulf and an absolute one, between the ruling and the serving classes. The two functions are opposed. And the function determines the individual.' For Clifford, a person's position was more important than who they were or what they did.

'Aristocracy is a function,' he said, 'a part of fate. And the masses are a functioning of another part of fate. The individual hardly matters.'

D.H. Lawrence understood the change that occurred around the First World War in a way that suddenly irrelevant aristocrats never could. For all that the novel was portrayed as a threat to the social order due to its sexual frankness, the real threat came from its discussion of the upper classes' inability to comprehend they were finished. Many novels attempted to come to terms with the irreversible change that occurred to the British ruling classes after the First World War, from Ford Madox Ford's *Parade's End* (1928) to L.P. Hartley's *The Go-Between* (1953), but they were not as brutally blunt as *Lady Chatterley's Lover*. The aristocracy's attempts to carry on as before, in Lawrence's eyes, made them into a form of zombie. They may have physically existed and were still moving, but they were quite dead.

In order to escape the living death of life with her impotent aristocrat husband, Lady Chatterley begins an affair with the gamekeeper, Oliver Mellors. Despite Lawrence's frankness and the socially taboo nature of the relationship, the emotional heart of the affair was not dissimilar to the ideal Christian union described at length in the books of Marie Stopes. Lady Chatterley needed to be fulfilled sexually in order to become physically, emotionally and spiritually alive, just as Stopes advised. That fulfilment could only come by the shared willingness of her and Mellors to give themselves to each other entirely and unconditionally. Giving up their individuality allowed the pair to achieve a sense of union akin to the ideal spiritual love that was the focus of so much of Stopes's poetry. Stopes would have been appalled by their marital status, but she would have recognised that the relationship between Chatterley and Mellors was loving, tender and emotionally intelligent – in a way that the coming sexual revolution would not be.

The idealised spiritual union promoted by Stopes and Lawrence was no match for the incoming tide of individualism. A more typical attitude to loosening sexual mores can be seen in the work of the American novelist Henry Miller, whose first novel *Tropic of*

Cancer (1934) was, as mentioned earlier, influenced by the sexual openness of Dalí and Buñuel's surrealist film *L'Âge d'or*. This semi-autobiographical modernist novel records the aimless life of Miller as he drifts penniless around Paris, failing to write his great novel. Like *Ulysses* and *Lady Chatterley's Lover*, it was the subject of numerous obscenity charges before a 1964 US Supreme Court decision found that it had literary merit and could be freely published.

Tropic of Cancer is a link between the novels of the early modernists and the Beat writers and existentialists to come. While its stream-of-consciousness approach and lack of interest in plot recall Joyce, the nihilism and self-centredness of the main character is emotionally closer to Sartre or Kerouac. Miller writes from a deeply misanthropic perspective. 'People are like lice,' he announces early in the book. As Anaïs Nin explains in the book's preface, 'Here is a book which, if such a thing were possible, might restore our appetite for the fundamental realities. The predominant note will seem one of bitterness, and bitterness there is, to the full. But there is also a wild extravagance, a mad gaiety, a verve, a gusto, at times almost a delirium.' It was this delirium that made *Tropic of Cancer* an important book, especially in the eyes of the Beats. But, as Nin warns, the overriding tone is cold-hearted.

Despite occasional epiphanic moments, such as one triggered by the lack of self-consciousness of a Parisian prostitute, Miller has no interest in any romanticised notion of spiritual union. The sexual encounters he details are motivated more by anger and disgust than by love and affection. Lady Chatterley would have been deeply unimpressed by Miller's performance as a lover. After having sex with his host's maid Elsa, Miller remarks that 'Somehow I feel sorry as hell for her and yet I don't give a damn.' It is a line with all the emotional intelligence and lack of self-awareness of a teenage boy. For Henry Miller, sex was about what he wanted. The needs of the other party were of little consequence. He originally considered calling the novel *Crazy Cock*.

The sexual revolution that Miller wanted became mainstream with the arrival of the oral contraceptive pill in 1960, which made birth control easier and more reliable. Because the Swingin' Sixties celebrated free love and is associated with great strides in civil rights, gay rights, vegetarianism and environmentalism, it is often assumed that it was also a period of female liberation, but this was not the case.

The feminist movement of the 1970s was necessary in part because of how women were treated in the 1960s. Women were significant players in the hippy movement and enthusiastic supporters of the relaxed sexual climate, but they were largely viewed, by both genders, as being in a supporting role to their men. Women who were following their own path, such as the Japanese artist Yoko Ono, were treated with suspicion.

In an era when any form of restraint on another's individuality was deeply uncool, men were quick to view women as objects to do with as they wished. As Bob Weir of the Grateful Dead reassured us in their 1971 song 'Jack Straw', 'We can share the women, we can share the wine.' Or, as Mungo Jerry sang in their 1970 pro-drink-driving anthem 'In the Summertime', 'If her daddy's rich, take her out for a meal / If her daddy's poor, just do what you feel.'

Watching British television from the 1960s and 1970s reveals the extent to which the objectification of women became normalised, at least in the minds of the producers of comedy and light-entertainment programming. A common trope was an older amorous man running after one or more younger women in a prolonged chase. This was considered funny, even though the fact that the women were running away demonstrated fear and a lack of consent. An example of this can be seen in a 1965 edition of the BBC family science fiction programme *Doctor Who*. The time-travelling Doctor (William Hartnell) was in ancient Rome where he witnessed Emperor Nero chasing a woman he intended to rape. The Doctor, not realising that the woman was his companion Barbara, smirks and waves his hand to dismiss the incident. The 'chasing women' trope became so normalised that comedians like Benny Hill were able to subvert it by having young women chase the old man. This didn't alter the fact that young women in Hill's programmes rarely spoke, and existed only to be ogled, groped and to undress.

Perhaps the nadir of the early 1970s objectification of women was the song 'Rape' by Peter Wyngarde. Wyngarde was a famous actor, best known for his portrayal of the womanising spy Jason King. With his bouffant hair, large moustache and flamboyant clothes, Jason King was a key inspiration for Mike Myers's comedy character Austin Powers. Wyngarde signed to RCA records and released an album in 1970. This included a song where he suavely discussed the differing pleasures of rape that resulted from raping women of different ethnicities, over an easy-listening musical background and the sound of women screaming. This was a song released by a major record label and performed by a famous celebrity at the height of his fame. It highlights how different the period around 1970 was to the present day, and indeed to the rest of history.

What made that era's portrayal of women unique was that extreme objectification was placed front and centre in popular culture, while sexual abuse, though rife, was hidden. Old hierarchical power structures meant that powerful men could abuse their positions with little danger of prosecution or public censure.

In the early 1970s many power structures from previous centuries were still in place, but they now existed in an individualist culture where women could be painted as voiceless objects. In this atmosphere, and within these structures, extreme institutionalised networks of sexual abusers were able to flourish. The extent of organised abuse of children within the British establishment is only slowly coming to light, but the open existence of the campaigning organisation Paedophile Information Exchange, which was founded in 1974 and received funding from the Home Office, gives some indication of the situation. Child abuse on a horrific scale by members of the Catholic Church was prevalent in many countries, of which Ireland, the United States and Canada have done the most to publicly investigate this dark history.

These institutionally protected networks of child abusers, clearly, had little concern for consent. They were not interested in the impact they had on their victims. The sexual life they pursued was a long way from the spiritual union sought by Marie Stopes or D.H. Lawrence.

These are extreme cases, but there is a pattern here. From the abusers in the Vatican and the British establishment to the attitudes of musicians and light entertainers, the sexual revolution during the 1960s was frequently understood through the perspective of individualism: get what you want, it was only your own concerns that mattered. People were finally free to live a physically fulfilled life, but focusing on the self caused many to choose an isolated, soulless form of sex.

Of all the key feminist texts that appeared in this atmosphere, the one that had the greatest impact was *The Female Eunuch* by the Australian academic Germaine Greer. Published in 1970, it has sold millions of copies and was translated into eleven languages. A scattershot mix of polemic and academic research delivered with more humour and plain speaking than many other feminist texts, it found an eager audience and became an immediate bestseller.

Greer recognised that the direction the sexual revolution was taking was not in the interests of women. 'Sex must be rescued from the traffic between powerful and powerless, masterful and mastered, sexual and neutral, to become a form of communication between potent, gentle, tender people,' she wrote. The alternative was the empty sexuality of the age, where 'we are never more uncommunicative, never more alone.'

The title of the book recognised that even though women had never been more objectified, they were not seen as fully sexual objects. They were like Barbie dolls, expected to be pretty and passive, but not possessing any genitals of their own. 'The female is considered as a sexual object for the use and appreciation of other sexual beings,' Greer wrote. 'Her sexuality is both denied and misrepresented by being identified as passivity. The vagina is obliterated from the imagery of femininity.' Betty Friedan had similar concerns. 'Sexual liberation is a misnomer if it denies the personhood of women,' she said. 'The first wave of so-called sexual liberation in

America, where women were passive sex objects, was not real liberation. For real liberation to be enjoyed by men and women, neither can be reduced to a passive role.'

Greer argued that the way forward for women was to recognise their innate self-worth and become fully sexual creatures. This would grant women 'freedom from being the thing looked at'. It would also, she noted, be a great gift to men.

The female liberation movement, which had been ignited by Friedan's *The Feminine Mystique*, was kicked into the mainstream by the success of Greer's book and became known as Women's Lib. Feminism had been on the back burner since the arrival of universal suffrage, but votes for women had not proved to be the magic bullet that the first wave of feminists had hoped. The ability to cast one vote every four or five years turned out to be a blunt tool for dealing with complex institutionalised bias. Gender equality in many areas, particularly pay equality, was stubbornly refusing to materialise. In 2015, this is still the case, but Women's Lib did make significant strides in many areas. The female objectification of the 1970s would not be accepted now.

Writers like these remind us that our culture is not as sexualised as it prides itself on being. The emotional intelligence needed for the individuality-shattering, full and committed relationships argued for by Stopes, Lawrence and Greer is frequently absent. For all the tits on display, a culture without communion will always be more masturbatory than sexual. When Philip Larkin wrote that 'Sexual intercourse began in nineteen sixty-three (which was rather late for me)', he may have been mistaken. It's possible that, on a cultural level, we're still waiting.

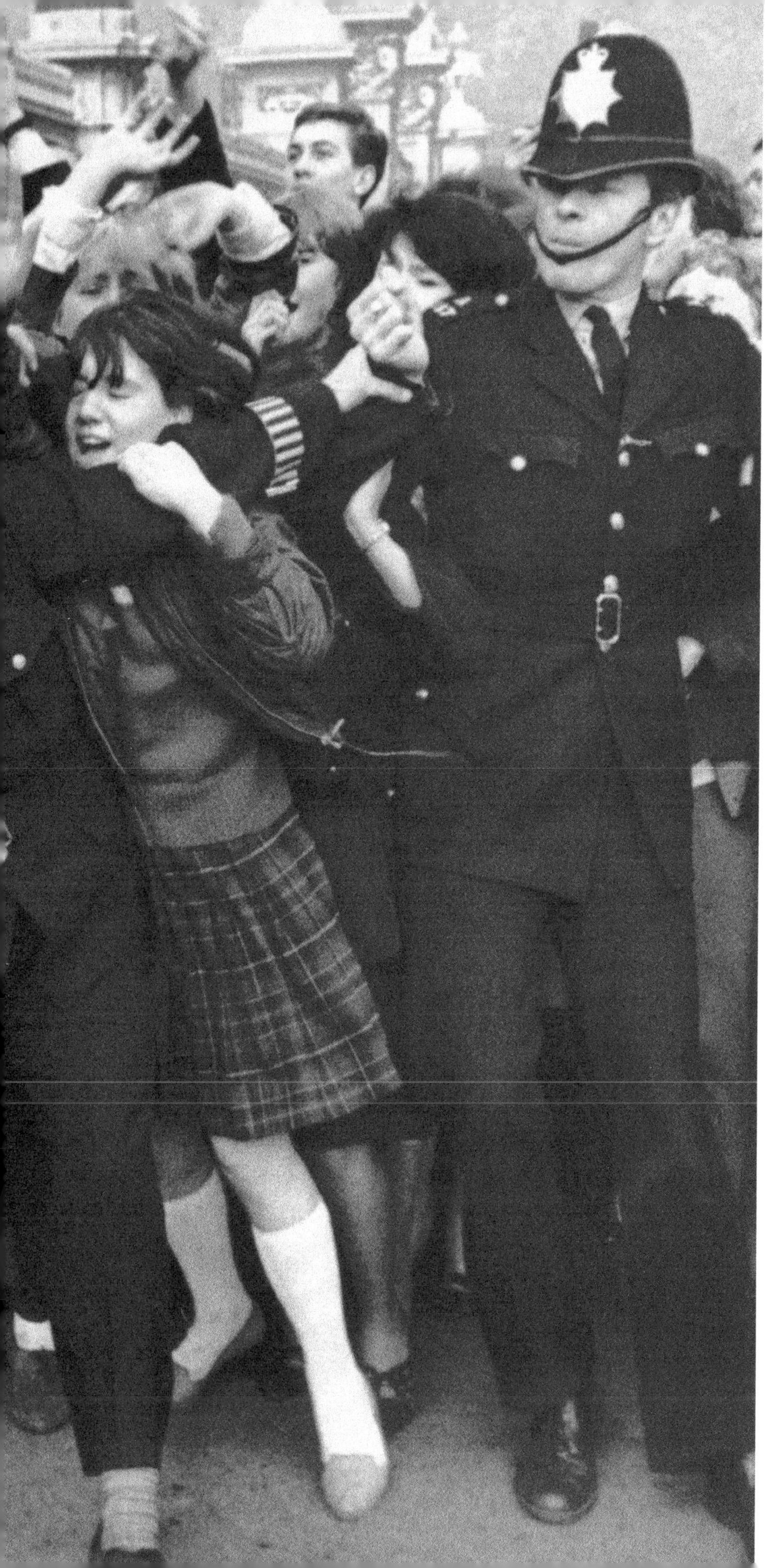

ELEVEN: TEENAGERS

Young fans outside Buckingham Palace as The Beatles receive their MBEs, 1965 (Hulton-Deutsch Collection/Corbis)

Wop-bom-a-loo-mop-a-lomp-bom-bom

Little Richard's 1955 single 'Tutti Frutti' began with a cry of 'Wop-bom-a-loo-mop-a-lomp-bom-bom!' Then came drums and twin saxophones, and the hammering of a piano. Little Richard was a twenty-five-year-old dishwasher from a poor town in Georgia, but on that song he announced himself, all hair and attitude, as a force of nature. Who had ever sounded that alive before?

It was a cultural year zero. Nonsense words were commonplace in music, but Little Richard screaming 'Wop-bom-a-loo-mop-a-lomp-bom-bom!' was entirely different to Perry Como singing 'Bibbidi-Bobbidi-Boo'. The record itself lasted little over two minutes, but its impact still echoes. When *MOJO* magazine produced a 'Top 100 Records that Changed the World' chart, 'Tutti Frutti' had to be number one.

Rock 'n' Roll had been slowly developing over a number of decades, with its roots reaching deep into the rhythm and blues, jazz, country and blues music of the southern United States. As a cultural phenomenon it was a product of the perfect storm of teenage energy that occurred in the mid-1950s. It was music aimed directly at white teenagers with disposable income and a desire to cut loose and have a good time. It fuelled the adoption of the electric guitar and the new seven-inch 45 rpm vinyl single format, which RCA introduced in 1949.

Television censors insisted that Elvis could only be filmed from the waist up because the mere sight of him was overtly sexual. 'Rock 'n' Roll [is] insistent savagery,' declared the 1956 *Encyclopaedia Britannica Yearbook*, 'deliberately competing with the artistic ideals of the jungle.' The name 'Rock 'n' Roll' itself was first used, in the early twentieth century, both as a euphemism for sex and also as a description of the waves of spiritual fervour in black gospel churches.

The music itself was an expression of ecstasy and as such was simultaneously sexual and spiritual. Its sexual nature was apparent in its lyrics. The first thing we are told about the eponymous heroine of Little Richard's 'Good Golly, Miss Molly' is that 'she sure love to ball'. But it was also evident beyond the lyrics, lurking in the music itself. The FBI undertook a two-year investigation into The Kingsmen's rock standard 'Louie Louie'. They attempted to decipher the song's slurred but innocent lyrics. The FBI's prosecutor LeRoy New eventually concluded that the lyrics were fine, but the music itself was lascivious and filthy. He described the song as 'an abomination of out-of-tune guitars, an overbearing jungle rhythm and clanging cymbals', but admitted that the obscenity laws 'just didn't reckon with dirty sounds'.

The best records generated a sense of direct revelation and communion, not filtered through the restricting middle man of the Church, and recognised the teenager's ever-present hormonal lust. Rock 'n' Roll from the 1950s, and the rock music that followed, generated a sense of undefined joy and a desire to share that experience with your sweetheart.

Little Richard was a cross-dressing, makeup-wearing bisexual black guy, and something of a challenge for the older conservative generation. He did at least make some attempt to tone down the sexual nature of his work. A discarded early draft of the lyrics for 'Tutti Frutti', for example, included the helpful advice, 'Tutti Frutti, good booty / If it don't fit, don't force it / You can grease it, make it easy'. But Little Richard also had a spiritual side. He quit secular Rock 'n' Roll for the life of a pastor after being deeply affected by the sight of a bright red fireball streaking across the sky during a concert in Australia in 1957. That fireball was almost certainly the launch of Sputnik 1.

At the end of the 1950s, a string of unrelated events conspired to remove almost all of the first generation of Rock 'n' Roll stars from the stage. Little Richard defected to the Church, Elvis Presley enlisted in the US Army, Jerry Lee Lewis was engulfed in a scandal following his marriage to his thirteen-year-old cousin, and Chuck Berry was arrested for transporting a fourteen-year-old girl across state lines. His legal case would drag out over the next couple of years, resulting in his being jailed in 1962. More tragically, a light aircraft carrying Ritchie Valens, The Big Bopper and Buddy Holly crashed in Iowa on 3 February 1959, killing all three musicians and the pilot. This event would become immortalised in the Don McLean song 'American Pie' as 'the day the music died'.

This sudden loss of most of the major players in an artistic movement was an event without parallel, and it seemed that the Rock 'n' Roll 'fad' would never recover. But the circumstances which generated the demand for a new teenage culture had not changed, and the next generation of bands soon arrived to fill the void. The musicians of the 1960s were able to grow rock music in ways that would have been unthinkable in the 1950s. The unintended 'scorched earth' effect caused by the loss of the first generation left behind a very rich soil indeed.

Keith Richards, the lead guitarist of The Rolling Stones, begins his autobiography with an account of a run-in with Arkansas police in 1975. Richards admits he knew he shouldn't have risked driving through the Southern Bible Belt. There had been controversy about the Stones being granted American visas for their tour, not least because of Richards's record for drug-related offences. The band's lawyers had warned him that Southern cops were itching to bust him. Yet he chose to ignore the danger and set off in a brand-new yellow Chevrolet Impala, laden down with a quantity of dope, pills, peyote and coke which even he considered excessive.

Richards was pulled over and taken to a police station in the tiny town of Fordyce, where a legal standoff developed. As the media and lawyers gathered, the police petitioned a judge for the legal right to open a suitcase of cocaine they found in his trunk. During all this he did his best to discard, or divert attention away from, the drugs hidden in his hat, about his person, and behind the door panels of his car. Richards could have opened his book with an anecdote about the success and acclaim that The Rolling Stones achieved, or about his deep lifelong love of American blues. Instead, he set the tone for his story with this farcical collision with the conservative legal establishment.

As he makes clear, he had no need to travel with all those drugs. He had cleaned up for the tour and 'wasn't taking the heavy shit at the time'. It would have been far safer to keep any drugs he did want with the rest of the band's equipment. Yet despite knowing how dangerous it would be, and having no need to do so, he loaded the car up with illegal substances and drove across Arkansas.

In the 1955 movie *The Wild One*, the young biker played by Marlon Brando was asked, 'Hey Johnny, what are you rebelling against?' 'What have you got?' he replies. He was, like the title character of the James Dean movie released the same year, a rebel without a cause. Their attitude was encapsulated in the 1966 B-movie *The Wild Angels*, in a conversation between an older minister and a young outlaw played by Peter Fonda. 'Just what is it that you want to do?' asks the minister. 'We want to be free, we want to be free to do what we want to do,' Fonda replies. 'We want to be free to ride our machines without being hassled by the man. And we want to get loaded, and we want to have a good time. That's what we're going to do. We're going to have a good time. We're going to have a party.'

This was Richards's attitude, which he summed up concisely with the statement, 'We needed to do what we wanted to do.' This was also Freud's id speaking, loud and uninhibited, unconcerned with society or anyone else except itself. It might have been easy to intellectually argue against such an attitude, but that didn't change the fact that repercussion-free individualism *felt great*.

During their mid-Sixties and early-Seventies heyday, The Rolling Stones symbolised the spirit of rebellious youth. They were 'bad boys', an unwholesome dangerous group that stood in contrast to the family appeal of The Beatles. They personified unrepentant individualism on a collision course with the staid establishment, not least because of the high-profile court case that followed the arrests of band members Mick Jagger, Keith Richards and Brian Jones for drug offences. Keith Richards's reputation for taking excessive amounts of drugs without dropping dead made him a rock god. To their young audience in the 1960s, The Rolling Stones represented freedom without consequences.

A predominant theme in the songs of The Rolling Stones was desiring, demanding or wanting. Mick Jagger would sing of how 'You Can't Always Get What You Want', or about how '(I Can't Get No) Satisfaction'. Compare this to the music of The Beatles. The attitude of 'I want' is relatively rare in their songs. When it does appear, in songs such as 'I Wanna Hold Your Hand' or 'I Want to Tell You', it indicates desire for human contact rather than a simple demand. The Beatles song that was written for The Rolling Stones, 'I Wanna Be Your Man', was also a plea for human contact, but one expressed in a more straightforward, demanding attitude than other Beatles songs. As Richards notes, '[The Beatles] deliberately aimed [that song] at us. They're songwriters, they are trying to flog their songs, it's Tin Pan Alley, and they thought this song would suit us.'

The Rolling Stones were consciously acting in the tradition of the 'Do What Thou Wilt' philosophy of Aleister Crowley and Ayn Rand. They employed satanic imagery in songs like 'Sympathy for the Devil' and albums like *Their Satanic Majesties Request*. They are known for the unashamed commercialism of their high ticket prices, Mick Jagger's admiration of Margaret Thatcher, and the manner in which their lawyers went after the song-writing royalties for 'Bitter Sweet Symphony' by The Verve, on the grounds that the band had used a previously cleared Rolling Stones sample excessively. Musically the band remained resolutely conservative, sticking with Richards's beloved blues tradition rather than branching out in more experimental directions. Many people were surprised when William Rees-Mogg, the famously right-wing editor of *The Times*, came out in support of The Rolling Stones during their trial for drug offences in an editorial memorably entitled 'Who Breaks a Butterfly on a Wheel?' But Rees-Mogg and The Rolling Stones were not, perhaps, as politically different as they might first appear.

If the attitude of The Rolling Stones can be boiled down to 'I want', then what is the philosophy of The Beatles? This was most clearly expressed in their contribution to *Our World*, the first global television broadcast which was broadcast to 400 million people in twenty-six countries in 1967 to celebrate the launch of a string of communication satellites. The Beatles performed a new, specially written song called 'All You Need Is Love'. Their career began with songs like 'She Loves You' and 'Love Me Do', and ended with Paul McCartney singing about how the love you take is equal to the love you make.

If The Rolling Stones were about *I want*, then The Beatles were about *love*. This did not mean that The Beatles were against material wealth. Their attitude to money can be seen in the scathing song *Taxman*, or in the quote, commonly ascribed to Paul McCartney, about how he and John Lennon would sit down and purposefully 'write [themselves] a swimming pool'. But judging by the evidence of their recorded output, material things were a secondary concern in The Beatles' philosophy.

The idea that 'all you need is love' was a product of the hippy counterculture of the mid- to late 1960s, and was strongly influenced by the band's interest in LSD and other psychedelic drugs. LSD did not give the user anything, but it amplified what already existed. This was not always pleasurable, so it was necessary to ensure that the circumstances surrounding taking the drug, the so-called 'set and setting', were favourable and positive. But the hippies were prepared to take that risk, because the drug offered a way of looking at the world that they found inspiring and rewarding.

One problem with the drug was that users would afterwards find the perspective change they experienced frustratingly difficult to describe or explain. LSD was very different to drugs like cocaine or alcohol, which are isolating and reinforce individualism. Perhaps the only thing that could be said with some certainty about psychedelic awareness was that it was very different to the individualistic outlook of 'I want'. The Rolling Stones may have had a brief

LSD-inspired psychedelic period, which produced their 1967 single 'We Love You', but they followed this up with *Their Satanic Majesties Request* a few months later.

LSD caused the user to see themselves not just as a self-contained and isolated individual entity, but as an integral part of something bigger. But explaining what this bigger thing was proved problematic, and resulted in the hippies talking vaguely of 'connection' and of how 'everything was one'. In this they were similar to the modernists, attempting to find a language with which to communicate a new, wider perspective. The hippies turned to Eastern religions, which they learnt about from American Beats like Allen Ginsberg and English writers like Aldous Huxley and Alan Watts. They tried describing their experiences in co-opted Buddhist and Hindu terms, but none of those ancient metaphors were entirely satisfactory in the modern technological age. As vague and simplistic as it may have sounded, it was simpler to fall back on the most universal non-individualist emotion to describe their experience: love. It was for this reason that the 1967 flowering of the psychedelic culture became known as the Summer of Love.

The emotion of love is an act of personal identification with an external other, when the awareness of that person is so overwhelming that any illusion of separation between the two collapses. There is a reason why the biblical term for physical love was 'to know' someone. As such it is distinctly different from the isolating individualism so dominant in the rest of the century. What it is not, however, is an easily extendable organisational principle that can readily be applied to society as a whole.

Christianity had done its best to promote love during the previous two centuries. The Church ordered its followers to love, through commandments such as 'love thy neighbour', as if this was reasonable or possible. But ordering people to love was about as realistic as ordering people not to love. Love just doesn't work that way, and it doesn't inspire confidence that the Church seemed to think it did. It is noticeable that the more individualistic strains of American Christianity, which bucked the global trend of declining congregations, put less emphasis on that faith's original teachings about love and social justice.

The love culture of the hippies was brought low by the ego-fuelling cocaine culture of the 1970s and 80s. Attempts at describing a non-individualistic perspective were dismissed for being drug-induced, and therefore false. The hippies' stumbling attempts to describe their new awareness had been too vague and insubstantial to survive these attacks and they were written off as embarrassing failures by the punks. Yet slowly, over the decades that followed, many of their ideas seeped into the cultural mainstream.

One way to understand the twentieth century's embrace of individualism is to raise a child and wait until he or she becomes a teenager.

A younger child accepts their place in the family hierarchy, but as soon as they become a teenager their attention shrinks from the wider group and focuses on themselves. Every incident or conversation becomes filtered through the ever-present analysis of 'What about me?' Even the most loving and caring child will exhibit thoughtlessness and self-obsession. The concerns of others become minor factors in their thinking, and attempts to highlight this are dismissed by the catch-all argument, 'It's not fair.' There is a neurological basis for this change. Neuroscientists report that adolescents are more self-aware and self-reflective than prepubescent children.

Aleister Crowley may have been on to something when he declared that the patriarchal age was ending and that the 'third aeon' we were entering would be the age of the 'crowned and conquering child'. The growth of individualism in the twentieth century was strikingly similar to the teenage perspective.

The teenage behavioural shift should not be seen as simple rudeness or rebellion. The child's adult identity is forged during this adolescent stage, and the initial retreat inwards appears to be an important part of the process. But something about the culture of the mid- to late twentieth century chimed with this process in a way that hadn't happened previously. In part this was the result of demographics, as the postwar baby boom meant that there was a lot more of this generation than usual. Adolescents were foregrounded and, for the first time, named. 'Teenager' is a word that was first coined in the 1940s. Like the words 'genocide' or 'racism', it is surprising to find that this term did not exist earlier.

A postwar generation gap emerged. The older demographic of parents and grandparents had lived through the Second World War. They saw friends and family die, knew that everything they held dear hung in the balance, and could not look to the future with any degree of certainty. But the world was a vastly different place for their children, and they treated it as such. The long global economic boom, which lasted from the end of the Second World War to the 1970s, had begun. There were jobs for those who wanted them, and that brought money for cars and radios and consumer goods. Conscription into foreign wars such as Vietnam aside, nobody was shooting at postwar teenagers. They had no reason to worry where the next meal was coming from. This was particularly true in the United States, which had not been devastated by the war and did not need to rebuild. It was, instead, a vast landscape full of natural resources in the early stages of a golden age. Things could be fun, as the teenagers saw it, if only the old squares would just lighten up and get off their backs.

Teenagers gained a reputation for violence and juvenile delinquency. In the eyes of the older generation, they were concerned with personal gratification at the expense of any greater purpose. As the teenagers saw it, the older generation were *stuck in the past* and irrelevant to the modern age. They just *didn't understand.* As the hippies in the 1960s would say, 'Never trust anyone over thirty.' The dividing line between the generations was marked by an abrupt change in dress. For men, the suits, ties and hats that had been the male uniform for generations were replaced with the more relaxed T-shirts, jeans and sneakers. If you were to show a teenager of the late twentieth century a photograph of any of the modernists, they would dismiss them as a boring old fart. That modernist may have been far more rebellious, dangerous and wild than the teenager, but their neat hair and three-piece suit would have been sufficient reason to dismiss them. What did the old culture have to tell them about life in the second half of the twentieth century anyway?

The youth culture of the 1950s was the beginning of the growth of a counterculture. It defined itself not by what it was, but by what it was not, because its purpose was to be an alternative to the mainstream. Countercultures have existed throughout history, from the followers of Socrates to the Daoist and Sufi movements, but the highly individualistic nature of the period was the perfect ecosystem for them to grow, flourish and run riot.

The counterculture historian Ken Goffman notes that for all countercultures may be defined by their clashes with those in power, that conflict is not what they are really about. Countercultures, he says, 'seek primarily to live with as much freedom from constraints on individual creative will as possible, wherever and however it is possible to do so'.

Over a period of nearly forty years, starting from the mid-1950s, an outpouring of individual creativity caused the teenage counterculture to grow and mutate in thrilling and unexpected directions. Each new generation of teenagers wanted their own scene, radically different to that of their older siblings. New technology and new drugs powered a period of continual reinvention and innovation. Rock 'n' Roll was replaced by Pyschedelia, which was replaced by Punk, which was replaced by Rave. Genres of music including Disco, Hip Hop, Reggae and Heavy Metal sprang up and fed into the sense of potential that so characterised popular music in the late twentieth century. These countercultures grew and spread until they had replaced the staid musical culture that they were created to reject. As their teenage audiences grew up, Rock 'n' Roll became the mainstream.

During this period the move towards individualism became politically entrenched. Its dominant position was cemented by the rise of Margaret Thatcher in Britain in the late 1970s. This led to arguments against individualism being rejected or attacked according to the tribal logic of politics.

Thatcher outlined her philosophy in an interview with *Woman's*

Own magazine, published on Hallowe'en 1987. She said, 'I think we have gone through a period when too many children and people have been given to understand "I have a problem, it is the Government's job to cope with it!" or "I have a problem, I will go and get a grant to cope with it!" "I am homeless, the Government must house me!" and so they are casting their problems on society and who is society? There is no such thing! There are individual men and women and there are families and no government can do anything except through people and people look to themselves first.'

Unusually, her government made a later statement to the *Sunday Times* in order to clarify this point. 'All too often the ills of this country are passed off as those of society,' the statement began. 'Similarly, when action is required, society is called upon to act. But society as such does not exist except as a concept. Society is made up of people. It is people who have duties and beliefs and resolve. It is people who get things done. [Margaret Thatcher] prefers to think in terms of the acts of individuals and families as the real sinews of society rather than of society as an abstract concept. Her approach to society reflects her fundamental belief in personal responsibility and choice. To leave things to "society" is to run away from the real decisions, practical responsibility and effective action.'

Thatcher's focus on the primacy of the individual as the foundation of her thinking was perfectly in step with the youth movements of her time. The main difference between Thatcher and the young was that she justified her philosophy by stressing the importance of responsibility. At first this appears to mark a clear gulf between her and the consequence-free individualism of The Rolling Stones. But Thatcher was only talking about individual *personal* responsibility, not responsibility for others. Personal responsibility is about not needing help from anyone else, so is essentially the philosophy of individualism restated in slightly worthier clothes.

This highlights the schizoid dichotomy at the heart of the British counterculture. It viewed itself as being stridently anti-Thatcher. It was appalled by what it saw as a hate-filled madwoman exercising power without compassion for others. It argued for a more

Beatles-esque world, one where an individual's connection to something larger was recognised. Yet it also promoted a Stones-like glorification of individualism, which helped to push British society in a more Thatcherite direction. So entrenched did this outlook become that all subsequent prime ministers to date – Major, Blair, Brown and Cameron – have been Thatcherite in their policies, if not always in their words.

This dichotomy cuts both ways. Numerous right-wing commentators have tried to argue that the phrase 'no such thing as society' has been taken out of context. It should in no way be interpreted to mean that Thatcher was individualist or selfish like the young, they claim, or that she thought that there was no such thing as society.

The pre-Thatcher state had functioned on the understanding that there *was* such a thing as society. Governments on both sides of the Atlantic had tried to find a workable middle ground between the laissez-faire capitalism of the nineteenth century and the new state communism of Russia or China. They had had some success in this project, from President Roosevelt's New Deal of the 1930s to the establishment of the UK's welfare state during Prime Minister Attlee's postwar government. The results may not have been perfect, but they were better than the restricting homogeny of life in the communist East, or the poverty and inequality of Victorian Britain. They resulted in a stable society where democracy could flourish and the extremes of political totalitarianism were unable to gain a serious hold. What postwar youth culture was rebelling against may indeed have been dull, and boring, and square. It may well have been a terminal buzz kill. But politically and historically speaking, it really wasn't the worst.

Members of youth movements may have regarded themselves as rebels and revolutionaries, but they were no threat to the capitalist system. It did not matter if they rejected shoes for Adidas, or suits for punk bondage trousers, or shirts for Iron Maiden T-shirts. Capitalism was entirely untroubled over whether someone wished to buy a Barry Manilow record or a Sex Pistols album. It was more than happy to sell organic food, spiritual travel destinations and Che Guevara posters to even the staunchest anti-capitalist.

Any conflict between the counterculture and the establishment occurred on the cultural level only; it did not get in the way of business. The counterculture has always been entrepreneurial. The desire of people to define themselves through newer and cooler cultures, and the fear of being seen as uncool or out of date, helped fuel the growth of disposable consumerism. The counterculture may have claimed that it was a reaction to the evils of a consumerist society, but promoting the importance of defining individual identity through the new and the cool only intensified consumer culture.

This was the dilemma that faced Kurt Cobain, the singer in the American grunge band Nirvana, in the early 1990s. A rejection of mainstream consumerist values was evident in everything he did, from the music he made to the clothes he wore. Yet none of that troubled the music industry. His music was sold to millions, as if it were no different to the music of manufactured teen bands such as New Kids on the Block. Cobain's values were, to the industry, a selling point which increased the consumerism he was against. His concern about his increasing fame was already evident on Nirvana's breakthrough album *Nevermind*, which went on to sell over 30 million copies. On the single *In Bloom* he attacks members of his audience who liked to sing along but didn't understand what he was saying. By the time Nirvana released their next and final studio album *In Utero*, Cobain seemed defeated by this contradiction. The album opened with Cobain complaining that while teenage angst had paid off well, he was now bored and old, and it contained songs with titles such as 'Radio Friendly Unit Shifter'. Cobain committed suicide the following year. As he wrote in his suicide note, 'All the warnings from the punk rock 101 courses over the years, since my first introduction to the, shall we say, ethics involved with independence and the embracement of your community, has proven to be very true.'

Cobain failed to reconcile his underground anti-consumerism beliefs with the mainstream success of his music. The 'all you need is love' strand of counterculture thought was never able to mount a successful defence against 'I want' individualism. For how, exactly, could the difficult task of identifying with something larger than the self compete with the easy appeal of liberation, desire and the sheer fun of individualism? Was there a way of understanding ourselves that recognised and incorporated the appeal of individualism, but which also avoided the isolation and meaninglessness of that philosophy? Cobain, unfortunately, died believing that such a perspective was in no way possible.

The second half of the twentieth century was culturally defined by adolescent teenage individualism. But despite complaints about kids being ungrateful and selfish, the adolescent stage is a necessary rite of passage for those evolving from children to adults. Understanding the world through the excluding filter of 'What about me?' is, ultimately, just a phase.

The teenage stage is intense. It is wild and fun and violent and unhappy, often at the same time. But it does not last long. The 'Thatcher Delusion' was that individualism was an end goal, rather than a developmental stage. Teenagers do not remain teenagers for ever.

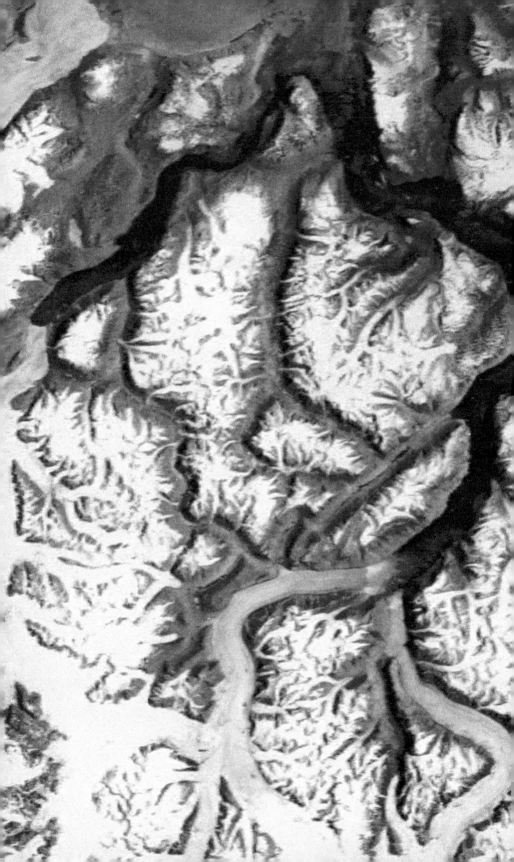

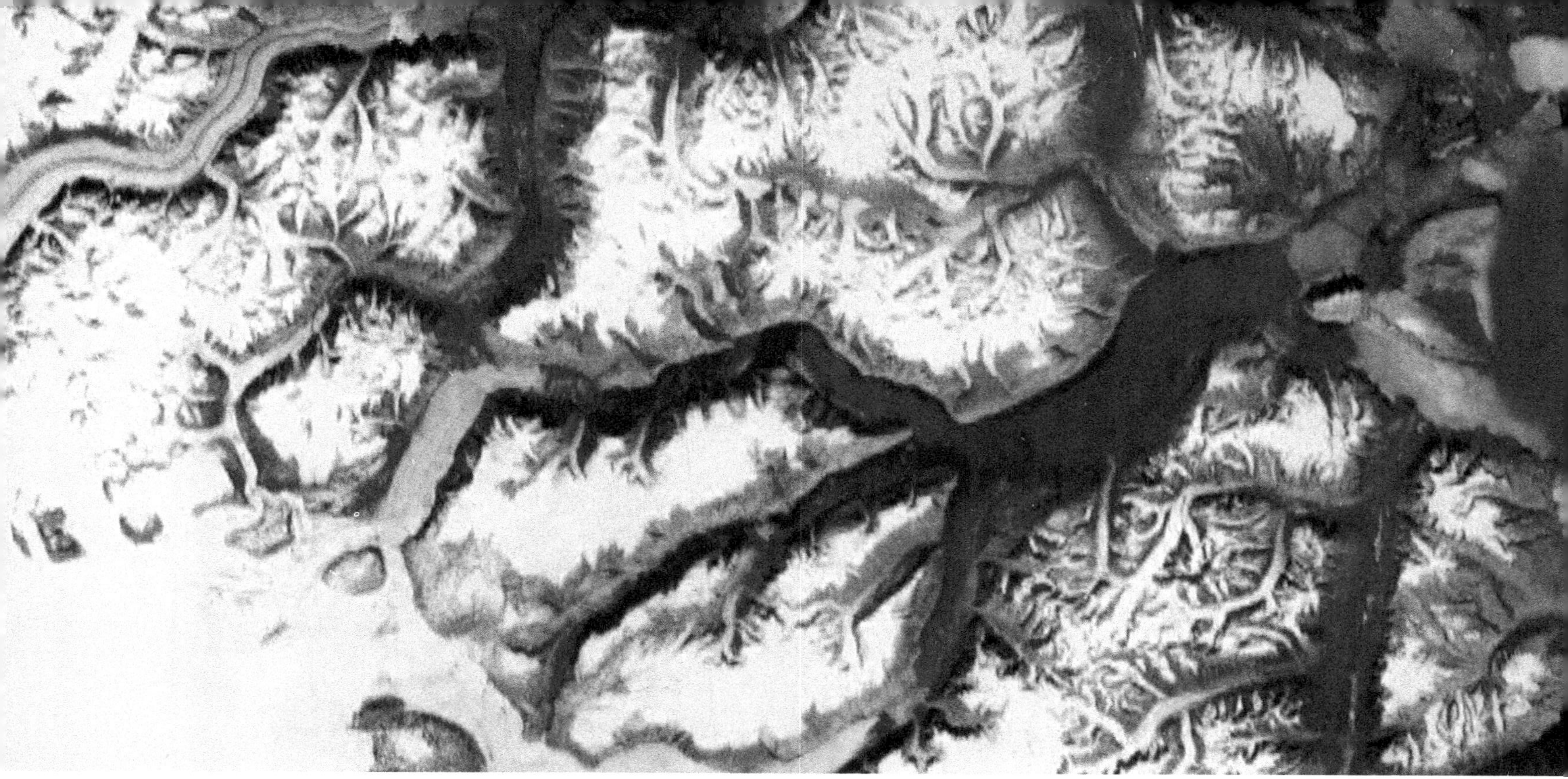

TWELVE: CHAOS

Fractal patterns formed by rivers in Greenland, photographed from space (Barcroft Media/Getty)

A butterfly flaps its wings in Tokyo

The universe, we used to think, was predictable.
We thought that it worked like a clockwork machine. After God had put it together, wound it up and switched it on, his job was done. He could relax somewhere, or behave in mysterious ways, because the universe would continue under its own steam. The events that occurred inside it would do so under strict natural laws. They would be preordained, in that they would transpire according to the inevitable process of cause and effect. If God were to switch the universe off and reset it back to its original starting state, then switch it back on again, it would repeat itself exactly. Anyone who knew how the universe worked and understood exactly what state it was in at any point would be able to work out what was coming next, and what would come after that, and so on.

That was not an idea that survived the twentieth century.

When Armstrong, Collins and Aldrin climbed into a tin can perched on top of a 111-metre-tall firecracker, it was Newton's laws which they were trusting to get them to the moon. In all credit to Newton, the laws he discovered over 250 years earlier did the job admirably. Relativity and quantum mechanics may have shown that his laws didn't work at the scale of the extremely small and the extremely large, but it still worked well for objects in between.

The mathematicians who performed the calculations needed to send Apollo 11 to the moon were aware that the figures they used would never be exact. They might proceed on the understanding that the total mass of the rocket was 2.8 million kilograms, or that the first-stage engines would burn for 150 seconds, or that the distance to the moon was 384,400 kilometres. These figures were accurate enough for their purposes, but they were always rough approximations. Even if those numbers were only out by a few hundred thousandths of a decimal place, they would still be out. But this wasn't a problem, because it was possible to compensate for any discrepancies between the maths and the actual voyage as the mission progressed. If the weight of the rocket was underestimated then it would travel a little faster than expected, or if the angle it left orbit at was slightly off then it would head increasingly off-course. Mission control or the astronauts themselves would then adjust their course by a quick blast of their steering rockets, and all would be well. This made complete philosophical sense to mathematicians. If the variables in their equations were slightly out it would affect the outcome, but in ways that were understandable and easily correctable.

That assumption lasted until 1960, when the American mathematician and meteorologist Edward Lorenz got hold of an early computer.

After he failed to convince President Eisenhower to unilaterally launch a nuclear attack on Russia, John von Neumann, the Budapest-born genius who inspired the character of Dr Strangelove, turned his attention to computers.

Von Neumann had a specific use in mind for computer technology. He believed that computer power would allow him to predict the weather, and also to control it. The weather, in his hands, would be a new form of 'ultimate weapon' which he would use to bury all of Russia under a new Ice Age. All the evidence suggests that von Neumann really didn't like Russia.

He became a pioneer in the world of computer architecture and programming. He designed an early computer that first ran in 1952 and which he called, in a possible moment of clarity, the MANIAC (an acronym for Mathematical Analyzer, Numerator, Integrator and Computer). He also designed the world's first computer virus in 1949. He was that type of guy.

His intentions for weather control beyond Russia were more altruistic. He wanted to trigger global warming by painting the polar ice caps purple. This would reduce the amount of sunlight that the ice reflected back into space, and hence warm the planet up nicely. Iceland could be as warm as Florida, he decided, which was fortunate because much of Florida itself would have been under water. Von Neumann, of course, didn't realise this. He just thought that a hotter planet would, on balance, be a welcome and positive thing. This idea was also expressed by the British Secretary of State for the Environment, Owen Paterson, in 2013. Von Neumann's thinking took place in the years before the discoveries of Edward Lorenz, so in his defence he cannot be said to be as crazy as Paterson.

Von Neumann died in 1957, so he did not live long enough to understand why he was wrong. Like many of the scientists involved in the development of America's nuclear weapon, he had been scornful of the idea that radiation exposure might be harmful. And also like many of those scientists, he died prematurely from an obscure form of cancer.

At the time, the idea of accurate weather prediction, and ultimately weather control, did not appear unreasonable. Plenty of natural systems were entirely predictable, from the height of the tides to the phases of the moon. These could be calculated with impressive accuracy using a few equations. The weather was more complicated than the tides, so it clearly needed more equations and more data to master it. This was where the new computing machines came in. With a machine to help with the extra maths involved, weather prediction looked like it should be eminently achievable. This was the reason why Edward Lorenz, a Connecticut-born mathematician who became a meteorologist while serving in the US Army Air Corps during the Second World War, sat down at an early computer and began modelling weather.

The machine was a Royal McBee, a mass of wires and vacuum tubes that was built by the Royal Typewriter Company of New York. It was a machine from the days before microprocessors and it would barely be recognisable as a computer to modern eyes, but it was sufficiently advanced for Lorenz to use it to model a simple weather system. His model did not include elements like rain, mist or mountains, but it was sophisticated enough to track the way the atmosphere moved around a perfectly spherical virtual planet.

Like the real weather, his virtual weather never repeated itself exactly. This was crucial, because if his weather conditions returned to the exact same state they had been in at an earlier point, then they would have started to repeat on a loop. His virtual weather would instantly become predictable in those circumstances, and real weather did not work like that. Yet as the constantly clattering output from his printer showed, his virtual weather did not loop. It was something of a surprise that such an unpredictable system could be recreated through a simple string of equations.

One day Lorenz decided to repeat a particularly interesting part of his weather model. He stopped the simulation and carefully reset all the variables to the state they had been in before the period he wanted to rerun. Then he set it going again and went to get a cup of coffee.

When he returned he found his weather system was doing something completely different to what it had done before. At first he thought he must have typed in one of the numbers wrong, but double-checking revealed that this was not the case. The model had started off mimicking its original run, but then the output had diverged. The difference was only slight to begin with, but it gradually increased until it was behaving in a way entirely unrelated to the original.

He eventually tracked the problem down to a rounding error. His machine held the numbers in its memory to an accuracy of six decimal places, but the numbers on the printout he had used when he reset the model were rounded down to three decimal places. It was the difference between a number such as 5.684219 and the number 5.684. It should not, in theory, have made much of a difference. If those numbers had been used to fire Apollo 11 at the moon, such a small difference would still have been accurate enough to send the spaceship in the correct general direction. Lorenz's weather was behaving as though the spacecraft had gone nowhere near the moon, and was performing an elaborate orbit around the sun instead.

This insight, that complex systems show sensitive dependence on initial conditions, was described in his 1963 paper *Deterministic Nonperiodic Flow*. This paper gave birth to a new field of study, commonly known as 'chaos mathematics'. In complicated systems such as the weather, minute variations in one variable could change the outcome in utterly unpredictable ways. Von Neumann's desire to master the weather would be, once this was understood, quite out of the question.

Lorenz popularised the idea through use of the phrase 'the butterfly effect'. If a single butterfly in Brazil decided to flap its wings, he explained, then that could ultimately decide whether a tornado formed in Texas. The butterfly effect does not mean that every flap of an insect's wings leads to tornados or other natural disasters; the circumstances which generate the potential for a tornado have to be in place. The point of the butterfly effect is that the question of whether that potential manifests or not can be traced back to a minute and seemingly irrelevant change in the system at an earlier point.

The idea of the butterfly effect appears in a 1952 short story by the American science fiction author Ray Bradbury called 'A Sound of Thunder'. In this story hunters from the future go back in time to hunt dinosaurs, but they must be careful to stay on levitating platforms and only kill animals who were about to die, in order not to affect history. They return to the future to find it changed, and realise that the reason was a crushed butterfly on the sole of one of their boots.

Lorenz was surprised to see such an unpredictable outcome from what was, with all the best will in the world, an unrealistically simple model. He wondered how simple a system could be and still never repeat itself exactly. To his surprise, he discovered unpredictability could be found in a simple waterwheel. This was just a wheel with leaking buckets around its rim, which would turn under the force of gravity when water was poured into the top. It was so simple that it appeared it must be predictable. A mathematician, engineer or physicist would, then, have scoffed at the idea that they wouldn't be able to predict its future behaviour.

Unlike his computer weather, which used twelve variables, the waterwheel could be modelled with just three: the speed the water runs into the wheel, the speed that the water leaks out of the buckets, and the friction involved in turning the wheel.

Some aspects of the model's behaviour were indeed simple. If the amount of water that fell into the buckets was not sufficient to overcome friction, or if the water leaked out of the buckets quicker than it poured into them, then the wheel would not turn. If enough water fell into the top buckets to turn the wheel, and most of that water leaked away by the time they reached the bottom, then the wheel would turn regularly and reliably. This is the state you'll find in a well-built waterwheel on the side of a mill. These two different scenarios – turning nicely and remaining still – are two of the different patterns this simple system could fall into. But a third option, that of unpredictable chaotic movement, was also possible.

If the amount of water pouring into the buckets was increased, then the buckets still had water in them when they reached the bottom. This meant that they were heavier when they went up the other side. The weight of the buckets going back up, in these circumstances, competed with the weight of the buckets coming down. This could mean that the wheel would stop and change direction. If the water continued to flow then the wheel could change directions repeatedly, displaying genuinely chaotic and unpredictable behaviour, and never settle down into a predictable pattern.

The wheel would always be in one of three different states. It had to be either still, turning clockwise or turning anticlockwise. But the way in which it switched from one state to another was unpredictable and chaotic, and the conditions which caused it to change – or not – could be so similar as to appear almost identical. Systems like this are called 'strange attractors'. The system is attracted to being in a certain state, but the reasons that cause it to flip from one state to another are strange indeed.

Strange attractors exist in systems far more complicated than Lorenz's three-variable model. The atmosphere of planets is one example. The constantly moving patterns of the earth's atmosphere are one potential state, but there are others. Early, simple computer models of the earth's weather would often flip into what was called the 'snowball earth' scenario, in which the entire planet was covered in snow. This would cause it to reflect so much sunlight back into space that it could not warm up again. A 'dead' planet such as Mars was another possible scenario, as was a hellish boiling inferno like Venus.

Whenever early climate models flipped into these alternative states, the experiment was stopped and the software reset. It was a clear sign that the climate models needed to be improved. The real earth avoids these states, just as a waterwheel on a mill turns steadily and reliably. Our atmosphere is fuelled by the continuous arrival of just enough energy from the sun, just as a functioning waterwheel is fuelled by the right amount of water pouring onto it from a river. It would take a major shift in the underlying variables of the system to cause the atmosphere to flip from one of these states into another. Of course, a major shift in one of the underlying variables of our atmosphere has been happening since the start of the industrial revolution, which is why climate scientists are so worried.

History and politics provide us with many examples of complicated systems suddenly switching from one state into another, for reasons that no one saw coming and which keep academics arguing for centuries – the French Revolution, the fall of the Soviet Union in 1991 and the sudden collapse of the global imperial system around the First World War. Strange attractors allowed mathematicians, for the first time, to see this process unfurl. It was, they were surprised to realise, not some rare exception to the norm but an integral part of how complex systems behaved. This knowledge was not the blessing it might appear. Seeing how systems flipped from one state to another brought home just how fragile and uncontrollable complex systems were.

Thanks to the butterfly effect, climate modelling has proved to be exponentially more difficult than von Neumann expected. But the need for weather forecasting and longer-range predictions has not gone away, so climate modellers have worked hard. In the half-century since Lorenz first coded a virtual atmosphere, climate models have become massively more detailed and computationally intensive. They have to be run many times in order to obtain the statistical likelihood of their results. As the models improved, they became far less likely to flip into 'snowball earth' or other unlikely states. And in doing so they confirmed a finding that had shocked the pioneers of chaos mathematics. Whenever they studied complexity and looked deeply at the frothing, unpredictable turbulence, they found the strangest thing. They found the emergence of order. The fact that our ecosystem was so complicated was what was keeping it so stable.

Benoît Mandelbrot was a magpie-minded Polish mathematician with a round, kindly face and the type of personality that found everything intensely interesting. He was a Warsaw-born Jew who fled the approaching Nazis as a child, first to France and then to America. In 1958 he joined the IBM Thomas J. Watson Research Center in New York to undertake pure research. This allowed him to follow his curiosity wherever it might take him.

In 1979 he began feeding a short equation into a computer. Like Lorenz's waterwheel model, Mandelbrot's equation was incredibly simple. It was little more than a multiplication and an addition, the sort of mathematics that could have been attempted at any point in history. The reason why it hadn't was because the equation was iterative, and needed to be calculated millions of times. The answer that came out at the end was fed back into the start, at which point the equation was performed again, and again, and again. This was why Mandelbrot needed a computer. Even the shabbiest early computer was happy to run a simple bit of maths as many times as you liked.

Mandelbrot wanted to create a visual representation of his equation, so he performed the same iterative mathematics for every pixel on his computer display. The outcome would be one of two things. Either the number would become increasingly small, and ultimately head towards zero, at which point Mandelbrot would mark that pixel on the screen black. Or the number would become massive and race off towards infinity, in which case that pixel would be coloured. The choice of colour varied with the speed at which that number increased. The set of numbers that created this image became known as the *Mandelbrot Set*.

The result, after the equation had been applied to the whole screen, was an appealing black splodge in the centre of the screen with coloured, crinkly edges. It wasn't a circle, exactly, but it was satisfyingly plump. The shape resembled a cross between a ladybird and a peach, or a snowman on its side. It wasn't a shape that anyone had ever seen before, yet it felt strangely familiar.

It was when he looked closely at the edges of the Mandelbrot Set that things got interesting.

The edges of the shape weren't smooth. They were wrinkled and unpredictable, and sometimes bulged out to form another near-circle. Zooming in to them should have clarified their shape, but it only revealed more and more detail. The closer you looked, the more you found. There were swirls that looked like elephant trunks, and branching shapes that looked like ferns or leaves. It didn't matter how closely you dived in, the patterns kept coming. There were even miniature versions of the initial shape hidden deep within itself. But at no point did the patterns repeat themselves exactly. They were always entirely new.

Mandelbrot had discovered infinite complexity hidden in one short equation.

It might have been expected that such complexity should have been entirely random and discordant, but that was not the case. It was aesthetically appealing. Mathematicians are notoriously quick to describe whatever they are working on as 'beautiful', but for once they had a point. There was something very natural and harmonious about the imagery. They were nothing like the sort of images then associated with computer graphics. Instead, they resembled the natural world of leaves, rivers or snowflakes.

Mandelbrot coined the word *fractal* to describe what he had discovered. A fractal, he said, was a shape that revealed details at all scales. An example of this would be the coastline of an island. This will always remain crinkly, regardless of whether you are looking at headlands, or rocks, or individual pebbles on the shore. The smaller the scale, the more detail that emerges.

For this reason, measuring the length of a coastline is an arbitrary exercise entirely dependent on the amount of detail factored into the measurements. The length of the British coastline is 17,820 kilometres, according to the Ordnance Survey, yet the *CIA Factbook* reckons that it is 12,429 kilometres, or nearly a third shorter. Those measurements are entirely dependent on the scale from which they were taken. The figure is essentially meaningless without that context. The observed can, once again, only be understood if we include its relation to the observer.

Having found fractals on his computer, Mandelbrot looked again at the real world and realised they were everywhere. He saw them in the shapes of clouds, and in the whirls of rising cigarette smoke. They were in the branching of trees and the shape of leaves. They were in snowflakes, and ice crystals, and the shape of the human lungs. They described the distribution of blood vessels, and the path of a flowing river. At one point Mandelbrot was invited to give a talk at the Littauer Center at Harvard University, and was surprised to turn up and find what looked like one of his diagrams already on the blackboard. The diagram Mandelbrot intended to talk about depicted income variation, which was data in which he discerned fractal patterns. The Harvard diagram already on the board had nothing to do with income variation. It represented eight years of cotton prices. Yet that data had also produced remarkably similar fractal patterns.

Every time Mandelbrot stepped outside his front door and into the fractal landscape of nature, he was confronted by a world that now appeared entirely different to the one imagined by the mathematics of Euclid and Newton. A mountain may roughly be the shape of a pyramid, but only roughly. The classic Euclidian geometric shapes of spheres, cubes, cones and cylinders didn't actually exist in the natural world. There was no such thing as a straight line until mathematicians invented one. Reality was far messier than it had been given credit for. Like it or not, reality was fractal and chaotic.

Through the insights of people like Lorenz and Mandelbrot, and the arrival of brute computational power, a major shift occurred in our understanding of both mathematics and nature. As the research accumulated, two surprising facts became apparent. When you looked closely at what appeared to be order, you would find outbreaks of chaos around the edges, threatening to break free. And yet when you looked deeply into chaos, you would find the rhythms and patterns of order.

The discovery of order in chaos was of intense interest to biologists. The existence of complex life never really appeared to be in the spirit of the second law of thermodynamics, which stated that, in an isolated system, entropy would increase. Ordered things must fall apart, in other words, so it was odd that evolution kept generating increasingly intricate order. With the arrival of chaos mathematics, biologists now had a key that helped them study the way order spontaneously arose in nature. It helped them understand the natural rhythms of life, both in the internal biology of individual animals and in the larger ecosystems around them.

It was not long before someone applied this insight to the largest ecosystem we have: planet Earth itself, and all the life that exists on it.

In the 1960s the English polymath James Lovelock was working at NASA when they were preparing to launch unmanned probes to Mars. Lovelock's work was focused on studying the Martian atmosphere. It led to him inventing chlorofluorocarbon (CFC) detectors, which later proved invaluable when CFCs were discovered to be the cause of the growing hole in the ozone layer.

The atmosphere of a dead planet should be very different to that of a living planet like Earth, and so analysing the Martian atmosphere would prove to be a useful clue in determining whether or not there was life on Mars. As it turned out, the atmosphere of Mars was very close to a natural chemical equilibrium, being predominantly carbon dioxide with very little of the more interesting gases such as oxygen or methane, and this strongly suggested that Mars was a dead planet.

As Lovelock pondered the atmospheric differences between a living planet and a dead one, he became increasingly intrigued by the processes in which living organisms altered their atmosphere. This could occur in many different ways. For example, an increase in temperature would stimulate the growth of phytoplankton, which are tiny plants that live on the surface of the ocean and which excrete a compound called dimethyl sulphide. Those extra plants produced extra dimethyl sulphide, which entered the atmosphere and made it easier for clouds to form. Those extra clouds then reflected more of the sun's energy back into space, which had a cooling effect on the climate and helped reduce the amount of phytoplankton back towards their initial quantity. The whole system was a feedback loop, which constantly acted upon itself.

Wherever Lovelock looked, at all manner of chemical, biological, mineral and human processes, he found countless similar examples of feedback loops. Order was spontaneously being generated by chaos. The ecosystems of earth were unwittingly stabilising the very conditions they needed to survive.

The work of Lovelock and his colleague, the microbiologist Dr Lynn Margulis, led to Gaia Theory. They argued that our planet could be considered to be a self-regulating organism that altered its physical state in order to maintain the necessary conditions for life. Life on earth was generating the conditions needed for the existence of life on earth, in other words, and the magnitude of its complexity had created an extraordinary amount of stability. The planet was behaving like a living thing. If you damaged it, it would heal itself – assuming, of course, that the wound was not too severe. This was still a chaotic system, after all, and there was no reason why it could not be pushed out of its stable equilibrium into the chaotic states studied by Lorenz. This is the fear that keeps climate scientists awake at night: not gradual climate change, but runaway climate change, where the natural rhythms of the planet tip into unpredictable chaos and the goal of growing sufficient food to feed 7 billion people becomes impossible.

Gaia Theory was naturally controversial, especially among those unfamiliar with non-linear mathematics. Scientists of the calibre of the evolutionary biologist Richard Dawkins and the biochemist Ford Doolittle criticised it on the grounds that they could see no mechanism whereby individual natural selection could allow for environmental-level concerns. But stability seemed to be an emergent property of nature. It was just something that happened, in a similar way to how life was something that just happened to matter, or consciousness was something that just happened to life. This idea made many scientists deeply uneasy. They understood that Lovelock was employing a metaphor, and that he was saying that the planet was acting *like* it was a living thing. But few wanted to be drawn into the problem of defining the difference between a system that was acting like a living thing, and one that actually was one.

Lovelock's ideas gradually became accepted, but only after some very specific definitions were put in place. They are now studied under the name Earth Systems Science, which differentiates itself from Gaia Theory by being extremely clear that it in no way suggests that the planet is regulating itself *consciously*.

The name 'Gaia' had been suggested to Lovelock by the English novelist Sir William Golding, the author of *Lord of the Flies*, and it proved to be something of a double-edged sword. It helped popularise the idea among the general public, but it also scared away many members of the scientific community. 'Gaia' was the ancient Greek goddess of the earth, and scientists were cautious about anything that might encourage the idea that the earth was some form of conscious deity. This was a sensitive area because, as the twentieth century progressed, a lot of people were indeed coming to view the earth as some form of conscious deity.

One of the more surprising aspects of twentieth-century Western spirituality was the growth of a wide variety of broadly sympathetic practices that are most simply generalised as pagan or neopagan. Paganism is focused on respect for the natural world, which it views as both living and divine. A prominent example would be the Wiccan faith founded by the Crowley-inspired English witch Gerald Gardner, which has grown to the point that it is now recognised by organisations ranging from the US Armed Forces to the UK's Pagan Police Association. Gardner's faith is largely unique in British history, for while Britain has a long history of giving the rest of the world stories, inventions, sport and music, it had never before given it a religion.

For those familiar with the power and influence of organised religion, it was easy to be unimpressed by paganism and its grass-roots, distributed nature. In the eyes of religions which claim authority from a central divine text, the sheer unstructured and contradictory variety inherent in paganism, and its focus on personal experience, rob it of any credibility. It was obviously all just made up. In the eyes of the pagans, however, it was the 'People of the Book' who lacked credibility. The notion that a single text could function as an ultimate authority in a post-omphalos individualist society appeared naive in the extreme. Individual personal experience, to a pagan, was the only authority they needed. This was the clash between the hierarchical imperial mind, in which a person is defined by their service to a higher Lord who takes responsibility for their protection and threatens them with punishment, and the individualism of the world after Einstein and the First World War.

Resistance to the ideas of Lovelock, and to the discoveries of climate science in general, has been particularly strong among the American religious right. As previously noted, American Christianity did not undergo the church-emptying declines of European Christianity. Its perspective was similar to that of Mother Teresa, who in 1988 said, 'Why should we care about the Earth when our duty is to the poor and the sick among us? God will take care of the Earth.' Here God is the 'big man' of the tribe who offers protection in return for service. Accepting that the climate could spiral into chaos is accepting that such protection does not exist. This makes working to prevent climate change ideologically problematic.

Before Apollo, those who had imagined looking back at the earth assumed that they would see what their prejudices expected to see. In D.H. Lawrence's *Lady Chatterley's Lover*, the bitter gamekeeper Oliver Mellors wonders how he can escape the sense of doom he feels about mankind. Even travelling to 'the moon wouldn't be far enough, because even there you could look back and see the earth, dirty, beastly, unsavoury among all the stars, made foul by men.'

But the idea that the earth was a single, integrated entity grew in the public imagination after the crew of Apollo 8 returned with a photograph of earth taken from outside its orbit. Images such as this became a catalyst for the emergence of the environmental movement in the 1970s. It is unsurprising that this affected humanity on a religious level, especially given the sheer beauty of the planet. As the great American science writer Carl Sagan wrote, 'Our posturings, our imagined self-importance, the delusion that we have some privileged position in the Universe, are challenged by this point of pale light [the image of earth from space]. Our planet is a lonely speck in the great enveloping cosmic dark. In our obscurity, in all this vastness, there is no hint that help will come from elsewhere to save us from ourselves.'

The Apollo 12 astronaut Alan Bean said that 'I find it curious that I never heard any astronaut say that he wanted to go to the Moon so he would be able to look back and see the Earth. We all wanted to see what the Moon looked like close up. Yet, for most of us, the most memorable sight was not of the Moon, but of our beautiful blue and white home, moving majestically around the sun, all alone in infinite black space.' That sight profoundly changed Bean and many of the other Apollo astronauts. Bean became a painter and a number of other Apollo astronauts became religious or spiritual.

This shift in personal perspective, caused by seeing first-hand how frail, beautiful and alone the earth is, is known as the 'overview effect'. It was most noticeable in crew members who had no major responsibilities during the flight home, and who could spend the journey looking out of the window at their home planet. It was a view that, once seen, changed them for ever.

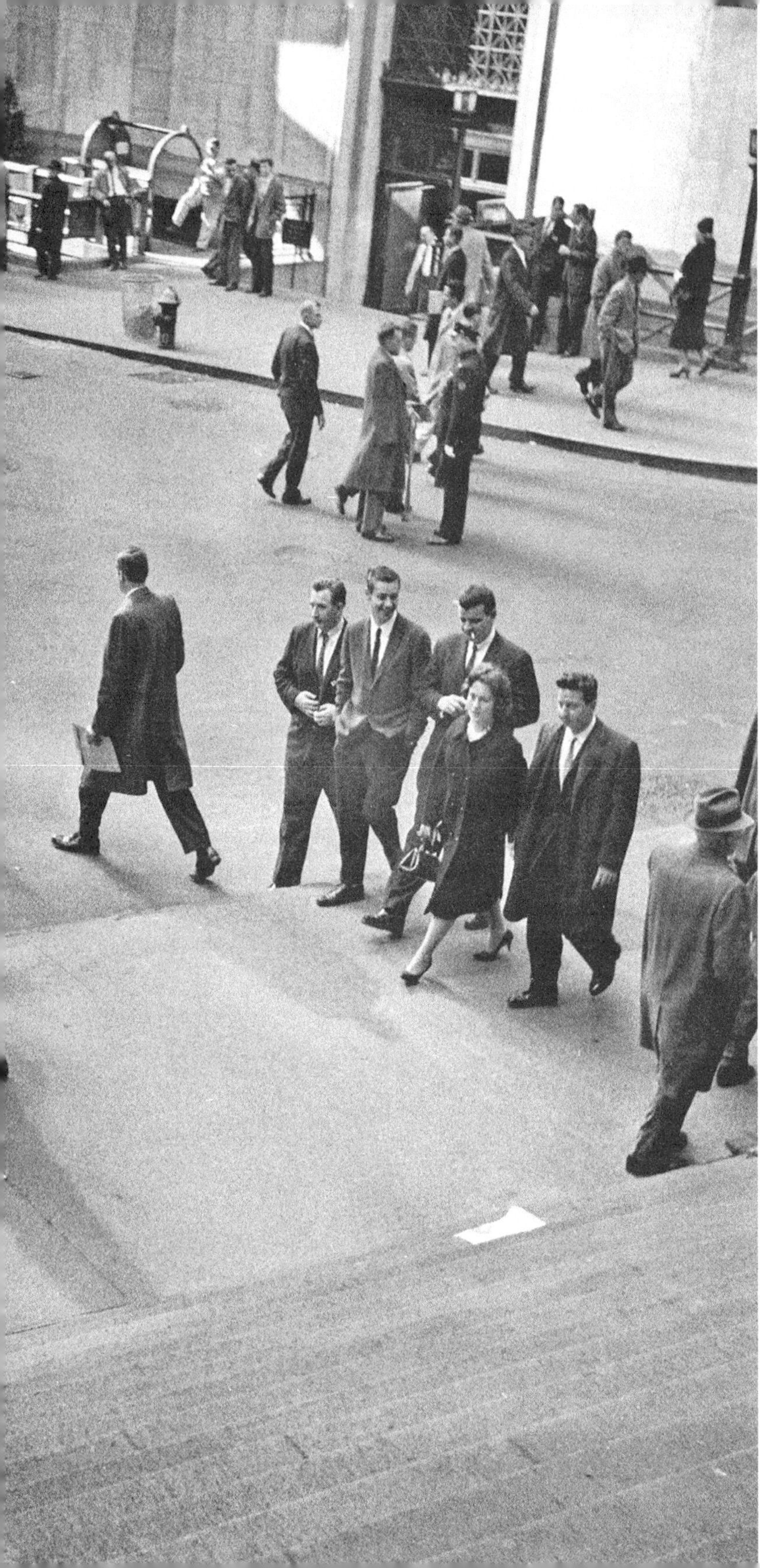

THIRTEEN: GROWTH

Wall Street, New York, c.1960 (Neil Libbert/Bridgeman)

Today's investor does not profit from yesterday's growth

The Golden Toad was a sociable critter, at least during mating season. Male Golden Toads were a couple of inches in length and congregated in large numbers on the mossy floor of the Costa Rican rainforest. They were a bright golden-orange colour. Female Golden Toads were a little larger and dark olive with small, yellow-ringed red blotches. Mating season had the air of a particularly decadent Day-Glo amphibian rave. When the American biologist Jay Savage first saw a Golden Toad he could not quite believe it was real, and wondered if someone had given it a coat of enamel paint.

We did not know the Golden Toad long. They were first discovered in the 1960s and were declared extinct in the 1990s. There are many other animals that became extinct in the twentieth century which we eulogise more, such as the Tasmanian Tiger, the Barbary Lion or the Baiji River Dolphin. The Golden Toads are simply one name on a very long list of plant and animal species that we'll never see again.

Estimates of the exact rate of species lost in the twentieth century range from the terrifying to the horrific. Extinctions are currently occurring at between 100 and 1,000 times the normal background level, and increasing. Primates, amphibians and tropical birds are being hit particularly hard. This is significant on a geological time-scale and has become known as the Sixth Extinction. To put it in context, the fifth great extinction occurred 65 million years ago, when a comet or asteroid collided with the earth and wiped out 75 per cent of plant and animal species, including the dinosaurs.

Some geologists now call the recent past the Anthropocene, a new geological term which recognises the extent to which the actions of human beings have impacted on the natural world. It is we who are responsible for this increased rate of extinction. There are a lot more of us now.

In 1900 the population of the world was 1.6 billion. Over the following century it quadrupled to over 6 billion, an unprecedented rate of increase. The economy also exploded. The Gross World Product, the sum of all economic activity in a particular year, was a little over a trillion dollars in 1900 but had grown to 41 trillion by 2000. The energy needed to fuel this growth increased by a factor of 10, with world energy consumption expanding from around 50 exajoules in 1900 to 500 exajoules by the end of the century. An exajoule is a quintillion joules, and a quintillion is even bigger than it sounds.

The sudden arrival of 4.6 billion people who need to consume an extra 450 quintillion joules and create 40 trillion dollars should really be the headline on any account of the twentieth century. Nothing else is remotely as significant.

For those familiar with chaos theory and the nature of complex systems, rates of change like this are truly alarming. Under normal conditions an intricate system such as our biosphere should tend towards stability. Exponential rates of change like these suggest that the normal system of self-limiting feedback loops has stopped working. Some factor was forcing our world into dangerous and unpredictable territory.

A good starting place to identify that factor would be the Fourteenth Amendment of the United States constitution.

The Fourteenth Amendment was ratified in 1868, following the Civil War. It stated that 'All persons born or naturalized in the United States, and subject to the jurisdiction thereof, are citizens of the United States and of the State wherein they reside. No State shall make or enforce any law which shall abridge the privileges or immunities of citizens of the United States; nor shall any State deprive any person of life, liberty, or property, without due process of law.'

The intention here was to ensure that the rights of former slaves became enshrined in law. In practice, the law was used very differently. The segregationist Jim Crow laws were one example of how the Amendment failed to achieve its aim. Of the 307 Fourteenth Amendment cases brought to court between 1890 and 1910, only 19 were brought by African Americans. The remaining 288 were brought by corporations.

Historically, corporations were organisations created by charters granted by monarchs or governments. These charters permitted a group of individuals to pursue a stated aim, such as building a school or a railway. These charters were restrictive, in comparison with partnerships or trusts, because the corporation's goals were limited and shareholders could be liable for the corporation's actions. But during the nineteenth century the shape of corporations began to change, and their power started to grow.

A key moment in this change was the landmark American case Trustees of Dartmouth College v. Woodward in 1819, which declared that the college's original charter, which had been granted by King George III of England in 1795, could not be overridden by the state of New Hampshire. This created a precedent which legally limited the power of states to interfere in the affairs of private corporations.

Following this decision, corporate lawyers began to look for ways to increase their powers further. What corporations really needed, or so those lawyers thought, was the type of freedoms guaranteed by the Fourteenth Amendment. When they looked at the wording of this amendment they saw that it granted rights to 'all persons'. And so, in one of those inspired flights of human imagination that suggest the involvement of alcohol, lawyers stepped into court and argued that those rights also applied to corporations, because a corporation was really a person.

There was actually some historical support for the idea. The legal status of monasteries had been awkward in the medieval period. They were often run by monks who had taken vows of personal poverty, and hence could not admit ownership or take legal responsibility for the monastery's infrastructure. To get round this headache, Pope Innocent V decided that the monasteries themselves could be considered as *legal persons*, and hence have some form of official existence. Crucially, he defined the term to specify that a legal person did not have a soul. By not having a soul, a legal person was exempt from the usual liability and ethical considerations that regular soul ownership conferred.

Corporate lawyers won their strange argument, and by the 1880s American courts, perhaps for the sake of a quiet life, routinely confirmed that corporations were individuals. They were, therefore, welcome to all the freedom and protection that the Fourteenth Amendment granted. It was a neat legal solution, and one which gave corporations the right to borrow money, buy and sell property, and take each other to court. But a corporation, plainly, wasn't an individual. An individual can be sent to jail for breaking the law. A corporation cannot, and the threat of imprisonment is an important limiting factor in the behaviour of individuals. Another difference is that people eventually die. A corporation could theoretically be immortal. Death is a necessary part of natural ecosystems, in part as a means of stopping things from growing too large.

Corporate shareholders had only limited liability for what the corporation did. They couldn't lose more than the value of their stock. This system made shares easy to trade and increased participation in stock markets, because the credit-worthiness of the stockholder was no longer a concern that the market had to take on board. Corporations, it seemed, had been granted the great dream of the twentieth century: individualism without responsibility.

But corporations did have a responsibility, and this strongly bound their actions. They were required by law to put their shareholders' interests above everything else, and that included the interests of their employees, their customers and the public good. Those shareholders' interests were almost entirely concerned with making money, so those corporations were legally bound to pursue maximum profit. Corporations had no choice but to become undying, unjailable profit-taking machines.

And as the wealth of corporations grew, financial quarter after financial quarter, so too did their power. Corporations started to catch up with, and then overtake, nation states. Of the one hundred largest economies in the year 2000, fifty-one were corporations and forty-nine were nations. Organisations such as General Motors, Exxon Mobil or Wal-Mart had become economically larger than countries like Norway, Saudi Arabia or Venezuela. And as their economic strength grew, so did their influence. Corporations took over the funding of politics and the media. There would be no attempts to control corporate growth from those directions.

The largest corporations demonstrate legal and financial superiority over nations. In December 2012 the US Justice Department was unable to prosecute the executives of the British bank HSBC on drug- and terrorism-related money-laundering charges, beyond a token fine. As Assistant Attorney General Lanny Breuer explained, 'Had the U.S. authorities decided to press criminal charges, HSBC would almost certainly have lost its banking license in the U.S., the future of the institution would have been under threat and the entire banking system would have been destabilized.' The corrupt bankers, in other words, were too powerful to be subject to the law of the most powerful nation on earth.

The reason why corporations were able to achieve such remarkable growth was, in part, due to externalities. An externality is the financial impact of an action which is not borne by the organisation that caused it. Externalities can be both positive and negative, but those created by organisations looking to reduce costs and increase profits tend to be negative. If a factory took so much water from a river that farmers downstream were no longer able to irrigate their land, then the farmers' financial losses would not be on the factory's balance sheet. Those losses, according to the factory's accounts, were an irrelevant externality. Other examples of externalities include noise pollution, systemic risk to the banking system caused by irresponsible speculation, and the emission of greenhouse gases.

An externality, in other words, is the gap between accountants and the real world. It is the severing of an important feedback loop, which could otherwise keep a complex system sustainable, ordered and self-stabilising.

One way to deal with the problem of externalities is through environmental laws and industrial regulation. But as the constant stream of corporate malfeasance cases shows, it is often more profitable to ignore these constraints and pay a fine should they get caught. The efforts of tobacco companies to cover up or cast doubt on the link between smoking and lung cancer, which was first established by the English epidemiologists Austin Bradford Hill and Richard Doll in 1950, was one example. Another was the refusal of Union Carbide to properly clean up after the 1984 Bhopal disaster, when thousands of people were killed by poison gas in the worst industrial accident in history. These are examples in which corporations chose to act in ways that would increase their profits but cause people to die. Pope Innocent V may have been showing great insight when he declared that an organisation with the status of a legal person did not possess a soul.

There are, unfortunately, a great number of similar examples. The actions of the Swiss corporation Nestlé highlight the failings of the legal system to prevent corporations knowingly causing deaths. In 1974 Nestlé was accused of encouraging malnutrition and infant deaths through its active promotion of formula baby milk to impoverished mothers in developing countries. Nestlé fought back by launching a libel trial against a War On Want report entitled, in its Swiss translation, *Nestlé Toten Babies* or Nestlé Kills Babies. A two-year trial followed. Despite the judge accepting the unethical nature of Nestlé's actions, he found in their favour on the grounds that the corporation could not be held responsible for the deaths 'in terms of criminal law'. The actions that led to those deaths were the legal responsibility of the corporate personhood of Nestlé itself, which could not be jailed. The individual executives who devised and implemented that behaviour could not be held individually liable.

Examples like these generate a great deal of anti-corporate sentiment. In their 2003 documentary *The Corporation*, filmmakers Mark Achbar and Jennifer Abbott examined the behaviour of corporations and asked, if a corporation really was a person, what sort of human personality did it exhibit? After cataloguing a lengthy list

of unethical behaviour including the inability to feel guilt, a failure to conform to social norms with respect to lawful behaviour
and the callous unconcern for the feelings of others, they came
to the conclusion that corporations should properly be classed as
psychopaths.

Corporations may not actually be individuals, but their pursuit
of their desires, and their rejection of related responsibility, is highly
reminiscent of teenagers from the 1950s onwards. To paraphrase
Peter Fonda, corporations just wanted to ride on their machines
without being hassled by the man, and that's what they were going
to do.

There are organisms in nature which depend on externalities
in order to grow and survive, such as parasites or cancer cells. But
these are typically found at the smaller scales of nature and, while
the cost of their externalities may be catastrophic for their hosts,
they can still be absorbed by the wider ecosystem. Behaviour that
can be absorbed on a small scale can have very different results
when it plays out on a larger stage.

The fact that corporations have grown to become such a large and
significant part of the world, while still being hardwired to pursue
perpetual growth, is deeply unnatural. Nature, in contrast, waxes
and wanes. It grows wild at times, but self-limiting feedback loops
always rein it back in again. And it is not just the natural world
that follows those rhythms. The development of passenger aeroplanes in the twentieth century saw them increase in speed until
the Anglo-French supersonic Concorde was flying from London to
New York in three hours and twenty minutes. But Concorde was
ultimately retired because it was expensive and inefficient, and the
same route now typically takes over seven hours.

The speed of commercial flight could not keep increasing indef-
initely. It was subject to economic and engineering factors, which
kept it within the boundaries required by a stable, functioning
transport system. Feedback loops rein in both the natural and man-
made world, keeping everything from the speed of planes to animal
populations within reasonable limits. But in a corporate economy

which actively promotes the pursuit of externalities, the natural feedback loops that would normally impede constant growth are severed, or simply ignored. Infinite economic growth can only exist by becoming divorced from reality.

For a typical Western individual in the middle of the twentieth century, all this was brilliant.

Economically, the first half of the twentieth century had been grim. It had been home to financial horror stories such as the hyperinflation of the Weimar Republic, where a glass of beer cost 4 billion marks in 1923, and the Great Depression itself. But the period from the end of the Second World War to the 1970s looks, from the perspective of the early twenty-first century, like something of a Golden Age. The immediate situation of the postwar world was bleak, and yet untold millions were lifted out of poverty over the decades that followed. Malnutrition and starvation in the Western world were mostly confined to the pages of history. Wages rose and people became acquainted with the term 'disposable income'. Regular working people gained access to everything from motor vehicles to central heating. Healthcare provision improved enormously, and life expectancy rose. The average man in England and Wales was expected to live to forty-six in 1900, but that figure rose by over a third to seventy-three by 1990. For women, the same figure rose from fifty to seventy-nine. In 1957 the British Prime Minister Harold Macmillan said, 'Let us be frank about it – most of our people have never had it so good. Go around the country, go to the industrial towns, go to the farms and you will see a state of prosperity such as we have never had in my lifetime – nor indeed in the history of this country.' From a contemporary perspective that statement seemed complacent, but from a historical viewpoint it was a fair comment.

The growth of corporations was a major factor in the rise in living standards. The President of General Motors, Charles Erwin Wilson, was appointed to the position of Secretary of Defense by President Eisenhower in 1953. This was, with hindsight, a clear example of corporate influence on government. When he was asked if his two roles represented a conflict of interest, he replied that he could not imagine a scenario that created a conflict 'because for years I thought what was good for the country was good for General Motors, and vice versa'. General Motors was then one of the largest employers in the world and would become the first American corporation to pay a tax bill of over a billion dollars. Its well-paid employees had disposable income which they spent on consumer goods, creating growth in other industries. This in turn created an affluent society full of ready customers who wanted to buy GM cars. Corporate growth had produced a virtuous circle, and society as a whole benefited.

This was a golden period for American industrial design. It would no longer be enough for consumer goods like chairs or toasters simply to be functional, when they could be both functional and aesthetically beautiful. A leading figure in this move was the Michigan-born designer Norman Bel Geddes, who started as a theatrical designer but then began to apply streamlining and aerodynamic principles to everyday objects. His work on the Futurama pavilion of the 1939 New York World Fair was immensely influential on the postwar generation of designers, such as the Paris-born Raymond Loewy, whose work included everything from oil company logos to Greyhound buses and Coca-Cola vending machines. Designers including Bel Geddes and Loewy dreamt up a visual language of America that was far superior to the visual language of the communist East. They exploited new materials like chrome and vinyl, and new methods of production such as moulds and stamping. Americans who watched *Flash Gordon* B-movies as children would grow up and buy cars with fantastic tailfins that deliberately echoed the rocket ships of optimistic science fiction. Consumers were made to keep spending through ideas like planned obsolescence, where products were designed to break early and need replacing. An example of this was the light bulb, whose life expectancy was reduced from around 2,500 hours to less than 1,000 by an illegal organisation known as the Phoebus Cartel, whose members included General Electric, Philips and Osram.

Advertising was no longer about informing rational consumers about the existence of products. It focused on defining those goods as necessary elements of a consumer's personal identity. Consumer goods aren't integral elements of personal identities, needless to say, but the advertising industry was not one to let truth interfere with persuasion.

We might forget names and faces and birthdays, but once jingles and slogans were in our heads they were there for good: Guinness is Good For You, Finger Lickin' Good, I Want My MTV, It's the Real Thing, Just Do It. Advertising was a form of black magic. It used the glamour of lifestyle, and an understanding of subconscious psychology, to take money from people in exchange for products they did not previously want or need. In his 2002 documentary *The Century of the Self*, the British filmmaker Adam Curtis highlighted the connections between the work of Sigmund Freud and the work of his great-grandson Matthew Freud, founder of the public relations firm Freud Communications. From this perspective, branding, marketing and public relations were arts that manipulated people's subconscious for financial gain, while at the same time convincing those being manipulated that not only were they in complete control, they were proudly expressing their own individualism.

The growth of individualism was, clearly, of immense benefit to corporations who needed their products to be consumed in large numbers. They did all they could to promote it.

It was an exciting time to be alive. A rising tide of affluence benefited entire populations and suggested that the future could only get better. The American Dream was the American reality. The mix of individualism, advertising and corporate growth was a potent cocktail indeed.

But then, at some point in the 1970s, things changed.

The strange attractor-like shift that occurred in the years leading up to 1980 was not apparent at the time. Economic growth continued as expected, but its impact on society began to change. The Princeton economist Paul Krugman calls the shift in American inequality that began at the end of the 1970s the 'Great Divergence', while the *New Yorker* journalist George Packer refers to the years after 1978 as the 'unwinding'. The benefits of economic growth slowly removed themselves from the broader middle class, and headed instead for the pockets of the very richest. Good jobs evaporated, social mobility declined and the 'millennial' generation born after 1980 are expected to be worse off than the postwar 'baby boomers'. Life expectancy rates have started to fall, in certain demographic groups at least. At the same time, inequality has increased to the point when, in 2015, the combined wealth of the eighty richest people in the world was equal to that of the poorest half of the world's population, some 3.5 billion people. Even those who believe that those eighty people work hard for their money would have difficulty arguing that they work 45 million times harder than everyone else.

This retreat of the American Dream, which had promised a future better than the past, is the result of a number of complicated and chaotically linked events from the 1970s. One of these was the rise of Deng Xiaoping to the position of Paramount Leader of the Chinese Communist Party in December 1978, in the aftermath of the death of Mao. Deng began the process of introducing a managed form of capitalism into China. The impacts of this would not be felt immediately, but the availability of cheaper Chinese labour for Western corporations would lead to the disappearance of well-paid Western manufacturing jobs, as well as destabilising trade imbalances. This process of globalisation also led to the disappearance of corporate taxes from government balance sheets. Corporations increasingly embraced globalisation and reimagined themselves as stateless entities in no way beholden to the nations that formed them.

A second factor was the collapse of the Bretton Woods Agreement in August 1971. This was a framework for international monetary management that had been agreed in the small New Hampshire town of Bretton Woods towards the end of the Second World War. The pre-war, every-man-for-himself approach to currency valuation had been responsible for some of the instability that led to war, so Bretton Woods was an attempt to create a more stable environment for international finance. It tied the value of international currencies to the US dollar, which was in turn tied to the value of gold reserves.

This link between the dollar and gold was important. When the Bank of England had first issued banknotes in the currency of 'pounds sterling' in the eighteenth century, this meant that the note itself was a convenient stand-in for a certain weight of sterling silver. The paper itself, therefore, was really worth something. For this reason, currencies have historically been backed by something that was both physical and in limited supply, such as gold or silver.

This link between money supply and actual physical wealth created confidence in a currency, but it was a constant frustration for those who dreamt of perpetual, continuous economic growth. The gold standard, as the link between paper money and precious metals was known, tended to produce periods of growth interspersed with periods of deflation and even depression. This may have been a natural and sustainable system, but it was not the sort of thing that democratic societies voted for. President Nixon's response to a period of economic pain was to end the gold standard, cut the link between the dollar and physical wealth, and ultimately bring Bretton Woods to an end. The value of the dollar could then float free, worth whatever the markets said it was worth.

Thanks to this neat trick of divorcing money from physical reality, the perpetual-growth economy continued in what would otherwise have been a period of recession. It also transpired that the ever-increasing amount of consumption needed for perpetual growth could be financed by creative bookkeeping, and the creation of debt. Debt mountains began to grow. When that approach ran into difficulty, in the early twenty-first century, taxpayer-funded bailouts kept the dream of perpetual growth alive.

Financial traders were able to create wealth out of thin air using wild and psychedelic financial instruments such as those traded on the derivatives market. This involved the trading not of actual things, but of changes to how the market would value things over time. It is no oversimplification to describe the derivatives market as pretty much incomprehensible, which is a problem for those who wish to regulate it. It was recently estimated to have a notional value of $700 trillion, or about ten times that of the entire global economy. In the opinion of the billionaire philanthropist Warren Buffett, 'derivatives are financial weapons of mass destruction, carrying dangers that, while now latent, are potentially lethal.'

The fact that markets such as these created wealth on paper, while not actually doing anything of value or creating anything tangible, did not unduly trouble those who profited from them. But when Adam Smith defined wealth in his 1776 book *The Wealth of Nations*, he said it was 'the annual produce of the land and labour of the society'. Economics was supposed to be a mathematical model of what happened in the real world.

Another factor in the emergence of the Great Divergence was the peak in US conventional oil production at the start of the 1970s, and the rise in the price of a barrel of oil from roughly $4 in 1970 to a price north of $100 in 2008. The price of oil has a proven impact upon national GDP, so even allowing for inflation, this was a significant extra cost which became a drag on economic growth.

This meant that corporations had to work harder to maintain the same rates of growth as before. As energy costs increased, so other costs needed to be reduced, and payroll and taxes were the most likely candidates. This in turn encouraged the move away from the virtuous circle that existed during the period of postwar economic growth. What was good for corporations, increasingly, was not what was good for countries.

The impact of the price of oil on the US economy became apparent in October 1973, when members of the Organisation of Petroleum Exporting Countries began a six-month embargo on sales to the US. This was a protest against the American rearmament of Israel in the aftermath of the Six-Day War. It caused an unprecedented peak in prices and shortages at the pumps for motorists. This did initially spur research into renewable energy, and resulted in the shift in car design towards more aerodynamic, but visually boring, vehicles. Yet the influence of oil corporations on American politics meant that the country ultimately committed itself to a hydrocarbon-based energy policy, regardless of the future cost. This was apparent when Reagan became President in 1981, and immediately ordered the removal of the solar panels President Carter had installed on the White House roof.

The intellectual justifications for the policies that led to the Great Divergence are commonly referred to by the term *neoliberalism*. Neoliberalism was a school of economic thought that dated back to the 1930s, but it only became the orthodox belief system for politicians and corporations following the election of Margaret Thatcher as the British prime minister in 1979 and the arrival of the economist Paul Volcker as Chairman of the US Federal Reserve in 1979. Neoliberalism, at its heart, argued that the state was just too dumb to be left in charge of people's wellbeing. It just didn't understand the nature of people in the way that the markets understood them. It had only a fraction of the information needed to make good decisions, and it was too slow, inept and politically motivated to put even that knowledge to good use.

As the neoliberalists saw it, the role of the state was to put in place a legal system that protected property rights and allowed for free trade and free markets, and to protect this system by military and police forces. State-owned industries needed to be placed in private ownership and run for profit. At that point the state had to step away and not interfere. Private enterprise would take care of the rest.

Neoliberalism was always going to create inequality, but it claimed that this was for the greater good. If a small elite became immensely wealthy, then this wealth would 'trickle down' to the rest of society. Or in a phrase which came to symbolise the 1980s, 'greed is good.' Wealth did not trickle down, needless to say. It passed upwards from the middle class to the very top. Few economists now take the idea of the trickle-down effect seriously, but the thinking behind it still colours much of the discussion about global economics. It is still common to hear the very rich described as 'wealth creators', rather than the more accurate 'wealth accumulators.'

The belief that a combination of free markets and property rights would solve all problems meant that sometimes it was necessary to create markets where they had not previously existed. It was blind faith in this logic that led, in 1997, to the World Bank pressuring Bolivia to grant foreign corporations ownership of all Bolivian water. This included rainwater that people had traditionally collected from their own roofs. According to the theory of neoliberalism, privately owned property rights such as these were the best way to give Bolivians access to water. The Bolivian people did not see it this way, especially after the corporations exploited their monopoly and immediately raised water prices by 35 per cent. The protests that followed led to martial law being declared, and one death, before the Bolivian people got their water back.

Neoliberalism produced a significant retreat in the role of the state, in comparison to the postwar Golden Age. When the world of empires was replaced by a world of nation states after the First World War, those nation states justified their new position by taking over the duty of protection which emperors had traditionally offered their subjects. This new form of *noblesse oblige* took the form of social and welfare programmes, as well as standing defence and police forces. This resulted in an increase in the size of the state. Total government spending in the US was under 10 per cent of GDP before the First World War, and typically between 30 and 35 per cent of GDP during the second half of the century. This pattern was repeated in most of the Western world. The reason why the neoliberals thought that they could reduce the size of the state was because corporations, as they grew to become more powerful than nations, were not beholden to any concept of *noblesse oblige*. It was simply not their job to protect ordinary people from psychopaths. Which was lucky, given the clinical assessment of their own personalities.

By the end of the twentieth century neoliberalism had become orthodoxy. As corporate power grew, its influence over politicians and media companies increased, in no small part because of their need for corporate money. Protests about corporate power occurred only outside of the political and cultural mainstream. The idea that a Western democratic politician from a mainstream political party could gain office with a platform that aimed to reduce corporate power, or increase corporate responsibility, became increasingly implausible. This is despite how popular a policy of, for example, making corporate executives legally responsible for their decisions would be with the electorate. There may have been widespread concern that a profit-led society was fundamentally inhuman, as well as depressing and unimaginative, but there was no way to express that opinion at the ballot box.

The years that followed the Great Divergence also produced the environmental movement. To the environmentalists, the neoliberalist pursuit of perpetual growth was delusional and deeply troubling. They had had their perspective on the earth radically changed by the photos brought back by the Apollo programme. While the planet had previously been imagined as an endless horizon and a bountiful, unexplored frontier, ripe for plunder, environmentalists now knew that it was a finite, closed system. The earth was limited, and in these circumstances the pursuit of perpetual growth was dangerous.

An old Indian legend illustrates the problem. An Indian king named Sharim was presented with an exquisitely made chessboard, and he was so delighted with the gift that he asked the craftsman to name the reward he would like in return. The craftsman asked for one grain of rice on the first square of the chessboard, two grains on the second, four grains on the third and so on, with the amount of rice doubling for each square on the board. King Sharim was surprised that the craftsman asked for such a trivial gift, and readily agreed. But the amount of rice increased rapidly from square to square. By the time they reached the twenty-first square it had reached over a million grains. By the last square the king was required to supply more rice than existed in the entire world. The amount of rice had not increased in the containable, linear fashion that the king had expected, but had instead increased geometrically. Geometric or exponential growth is like compound interest, in that the increases towards the end massively overshadow the increases at the start.

The doubling of rice is an extreme example, because those who believe in perpetual economic growth do not expect the economy to actually double each year. Yet even a seemingly manageable rate of growth, such as 2 per cent a year, requires the economy to double in size every thirty-five years. That means that twice as much real-world trade and economic activity have to occur in the space of roughly one human generation. But that is not the end of the problem because the amount of growth then continues to expand exponentially. It does not take long for the required size of the economy to become absurd.

All of this raises the question of when a global economic system based on perpetual growth will collide with the physical reality of a finite planet.

This was the question asked by the global think tank the Club of Rome, who in 1972 produced an influential book called *The Limits to Growth*. *The Limits to Growth* examined the implication of exponential growth in a number of categories, from human population to food production and resource depletion, and from that it generated a number of potential future scenarios. In one of those scenarios, the world stabilised in the mid- to late twenty-first century and fell into a sustainable system. In the other two scenarios, it didn't. The result was societal and economic collapse.

The Limits to Growth stressed that it was not attempting to make definitive predictions and was instead attempting to understand the behavioural tendencies of the global system. That said, a number of follow-up studies undertaken over thirty years after its publication have reported that the current data is roughly in line with its projections. Unfortunately those are the projections that point to overshoot and collapse, rather than the one that points to stabilisation. Increasing inequality of wealth seems to make the situation worse. It is the rich and powerful who are most able to change the system, but they are the last to be affected by collapse and have a greater investment in maintaining the status quo.

The reaction to *The Limits to Growth* was telling. It was rejected out of hand not by those who engaged with its data or arguments, but by those who were ideologically invested in the neoliberal project. It threatened constraints on individual behaviour and was dismissed for those reasons. It was not necessary to study the research into deforestation, depletion of topsoil, over-fishing or increasing water salinisation. The environmental perspective had to be wrong, because it was incompatible with individualism. Less than a century after our understanding of ourselves had been dominated by the top-down, hierarchical framework of masters and subjects, deferment to the desires of the individual had firmly cemented itself as our unshakeable new omphalos.

Environmentalism, from this perspective, was anti-human scaremongering which failed to take into account mankind's ingenuity. Imagination was a non-limited resource, and humans could adapt and solve problems as they developed. *The Limits to Growth*, it was argued, was no different to *An Essay on the Principle of Population* by the English clergyman Thomas Malthus. Writing at the end of the eighteenth century, Malthus had argued that population growth would lead to mass starvation. This scenario failed to materialise, in Western countries at least, thanks in part to the development of pesticides and fertilisers. But in a race against exponential growth it does not follow that, just because you keep up at the start, you will be able to keep up permanently. Exponential growth was like a video game which becomes increasingly difficult the more you play. The fact that you can complete level one doesn't mean that you will make it through level twenty-three in one piece.

The most significant, if unspoken, question about the alarms raised by environmentalists was this: would the point where the system collapses occur *after my lifetime*? For many of the baby-boomer generation, then comfortably into middle age, environmentalism didn't seem worth rejecting individualism for because the economic system looked like it should be able to keep going for at least another three or four decades. That baby boomers would think this with no regard for their children and grandchildren is a particularly damning indictment of individualism.

The clash between individualism and environmentalists is perhaps best illustrated by the global reaction to climate change. By the late 1980s it had become clear that the release of greenhouse gases on an industrial scale was affecting the climate in a manner which, if it continued, would be catastrophic. Crucially, there was still time to prevent this. The issue quickly reached the global stage thanks in part to influential speeches by Margaret Thatcher, most notably her 1989 address to the United Nations General Assembly. Thatcher was a trained chemist with a good grasp of the underlying science. 'The problem of global climate change is one that affects us all and action will only be effective if it is taken at the international level,' she said. 'It is no good squabbling over who is responsible or who should pay. Whole areas of our planet could be subject to drought and starvation if the pattern of rains and monsoons were to change as a result of the destruction of forests and the accumulation of greenhouse gases. We have to look forward not backward and we shall only succeed in dealing with the problems through a vast international, cooperative effort.'

This did not sit well with oil corporations. Selling hydrocarbons was a far easier way to achieve short-term profitability than a long-term research programme into alternative energy infrastructure. The technical challenges involved in producing carbon-free energy at a price and quantity that rivalled oil were, as scientists would say, 'non-trivial'.

The oil corporations and free-market think tanks began exercising their influence, in both government and the media, in an effort to prevent the international action on climate change that Thatcher spoke of. Their main tactic was a stalling approach which promoted a fictitious sense of doubt about the scientific consensus. This was an approach borrowed from the tobacco industry, which had used a similar disinformation campaign to cast doubt on the links between smoking and lung cancer. Those links were first discovered in 1950, but the tobacco industry was able to pretend otherwise for over four decades. Their campaign was highly successful in corporate terms because, even though hundreds of thousands of people died in one of the most unpleasant ways possible, they made loads of money and nobody went to jail.

In a similar way, the disinformation campaign of the oil industry was able to postpone action on climate change. It made it politically impossible for the United States to ratify the 1997 Kyoto Protocol, which aimed to set binding obligations on the reduction of greenhouse gas emissions from industrialised nations. After every typhoon, drought or flood, news programmes could be relied on to broadcast politicians angry at the suggestion that the extreme weather events now occurring could be linked to science which says that extreme weather events will increasingly occur. Even Margaret Thatcher had to amend her views after it became clear how much they offended her political allies. While her 1980 talks displayed clear scientific understanding of the situation, her 2003 book, *Statecraft*, fell back on the political talking points that cause climate scientists to bang their heads on their desks in despair. Curbing climate change was a front for a political viewpoint that she disagreed with, and for that reason no efforts to curb climate change should be made. Ideology beat science. Individualism beat environmentalism. So carbon continued to be emitted, topsoil continued to decrease and the ice sheets on the poles continued to melt. The debt which funded the consumer activity that caused all this continued to grow. As a result, the window when runaway climate change could have been prevented now appears to have closed.

And in the background, the Sixth Extinction continued. What chance did the Golden Toads have in a century such as that?

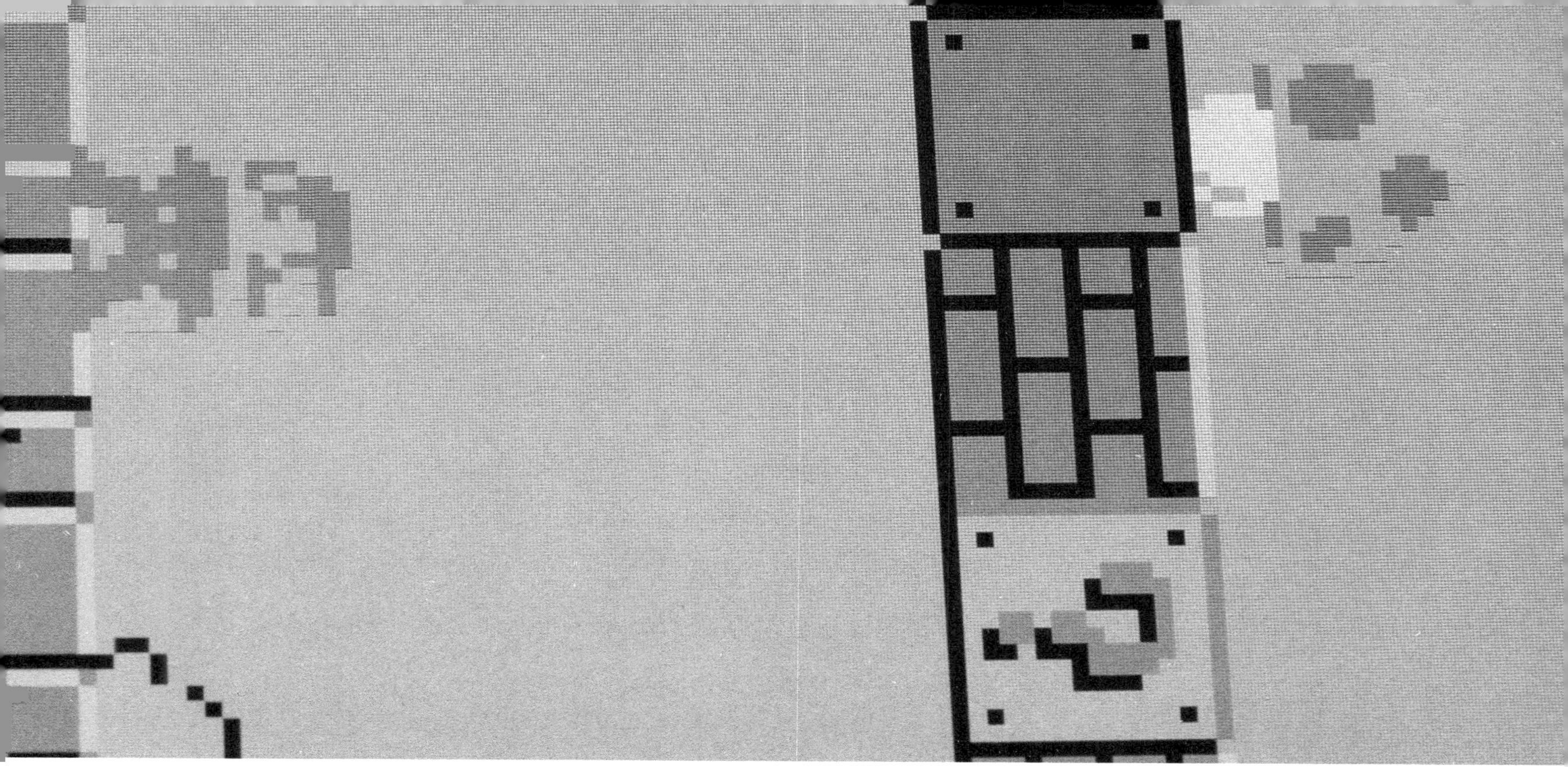

FOURTEEN: POSTMODERNISM

A screenshot from the 1985 Nintendo computer game Super Mario Bros. (ilbusca/ iStock)

I happen to have Mr McLuhan right here

If you want to understand postmodernism you should spend a few hours playing Super Mario Bros., a 1985 video game designed by Japan's Shigeru Miyamoto for the Nintendo Entertainment System.

In Super Mario Bros. the player takes control of a mustachioed Italian plumber named Mario. Mario's job is to travel across the Mushroom Kingdom in order to rescue Princess Peach, who has been kidnapped by Bowser, the monster-king of the turtle-like Koopa people. None of that, it is worth stressing, makes any sense.

Super Mario Bros. is a combination of elements that don't fit together under any system of categorisation, other than the game's own logic. Fantasy kingdoms are all well and good, but they are not usually the playground of Italian plumbers. Likewise the mix of elements Mario encounters in the game, from giant bullets to fire-spitting pot plants, does not lend itself to logical scrutiny. There is no need to look for hidden meaning in the symbolism of Super Mario Bros., because it isn't there. The character of Bowser, for example, was originally intended to be an ox, but he became a turtle-beast simply because Miyamoto's original drawing looked more like a turtle than an ox. Mario himself was also something of an accident. He originally appeared in the arcade game Donkey Kong and was known as Jumpman, because he was a man who could jump. He was later christened Mario as an in-joke, in honour of the landlord who owned the warehouse that was being rented by Nintendo of America. Princess Peach was rechristened Princess Toadstool for the American version of the game, for no reason of any importance.

None of these things affected the success of the game. What mattered was that each element was fun in itself. This is probably the most recognisable aspect of postmodernism, a collision of unrelated forms that are put together and expected to work on their own terms. The idea that an outside opinion or authority can declare that some elements belong together while others do not has been firmly rejected.

A related aspect of postmodernism is what theorists call *jouissance*. *Jouissance* refers to a sense of playfulness. The French word is used over its closest English translation, 'enjoyment', because it has a more transgressive and sexualised edge that the English word lacks. Postmodern art is delighted, rather than ashamed, by the fact that it has thrown together a bunch of disparate unconnected elements. It takes genuine pleasure in the fact that it has done something that it is not supposed to do. A good example of postmodern *jouissance* can be found in the British dance records from the late 1980s, such as MARRS' 'Pump Up The Volume' or 'Whitney Joins The JAMs' by The Justified Ancients of Mu Mu. These were records made by musicians who had just gained access to samplers and were exploring what they could do. They were having a whale of a time playing around and putting together all sorts of unconnected audio.

A third postmodern element can be seen in the mass-produced nature of the game. Super Mario Bros. is made from code, and that code is copied to create every instance of the game. It is not the case that there is one 'real' version of the game, while the rest are inferior imitations. The code that ran on Shigeru Miyamoto's development system, at the moment he signed the game off as complete, does not have some quality of authenticity that a battered second-hand copy found in a market in Utrecht does not. The status of identical copies of a work of art had been a hot topic in the art world ever since the German critic Walter Benjamin's 1936 essay *The Work of Art in the Age of Mechanical Reproduction*. As far as postmodernists were concerned, that debate was over. Every mass-produced copy of Super Mario Bros. was intrinsically as good as all the others, and no amount of hoping to find some magical aura imbued in an artist's own copy could change that.

A fourth important factor is that the game is well aware that it is a game. Super Mario Bros. makes no attempts to hide the conventions of the form, and will regularly highlight them in a way that games such as chess or tennis do not. Should the player find and collect a green and orange '1-Up' mushroom, they will be rewarded with an extra life and hence extend their playing time. In a similar way, the game is littered with rewards, power-ups and other gameplay factors that affect the structure of play, and which only make sense in the context of a video game.

This self-aware element of postmodernism is sometimes associated with film, television or theatre, such as the 1977 Woody Allen movie *Annie Hall*. Allen's character was able to win an argument with a self-righteous bore in a cinema queue by producing the media critic Marshall McLuhan from off-screen. At this point Allen turned to the camera and said, directly to the audience, 'Boy, if life were only like this!' In doing so he acknowledged the artificial nature of the situation: that he was a character in a movie, talking to a camera, in order to address a future audience of cinemagoers.

Postmodern moments like this are rare in the narrative arts, because they rely on the suspension of disbelief for their power. They are more common in the genre of comedy, such as the work of the British comedy troupe Monty Python. The final sequence in the 'Spanish Inquisition' episode of their second television series involved three members of the Spanish Inquisition being late for a sketch that they were supposed to appear in. Once this was realised they hurried off and caught a bus in order to get to the sketch. They knew that they were running out of time because the end credits had started rolling over them. They finally arrived in the sketch at the moment the programme ended.

Another postmodern aspect of Super Mario Bros. is that each time the game is played, it is different. There is no one true version of the game, and hence no true 'authorial intent' to provide the correct understanding of Miyamoto's work. Some users even go so far as to alter the code in order to create different versions of the game, known as mods. For gamers, this is entirely valid.

Postmodernists have firmly internalised Duchamp's insight that when different people read a book or watch a movie, they perceive it differently. There are many interpretations of a work, and it cannot justifiably be argued that one particular perspective is the 'true' one, even when that perspective is the author's. People can find value in a work by interpreting it in a way that the author had never thought of.

Finally, the game itself transcends the categories of highbrow and lowbrow, being simultaneously high art and populist fluff. When Super Mario Bros. was released in 1985 cultural critics would have dismissed it as lowbrow, had they been aware of it at all. Video games were then seen as dumb, noisy things for kids, and it took a number of decades before claims for their cultural validity were heard. Yet Super Mario Bros. was named as the best game of all time by IGN in 2005. It becomes difficult to classify a dumb bit of kids' entertainment, which is hailed as the pinnacle of a recognised art form, as being either highbrow or lowbrow.

Monty Python were a good example of the way postmodern culture was happy to be deep and shallow at the same time. Their 'Philosophers' Football Match' sketch depicted a game of football between German and Ancient Greek philosophers. Like much of their comedy, it was both silly and clever. As the football commentator describes the match, 'Hegel is arguing that the reality is merely an a priori adjunct of non-naturalistic ethics, Kant via the categorical imperative is holding that ontologically it exists only in the imagination, and Marx is claiming it was offside.'

Shigeru Miyamoto, it is not controversial to claim, is the most important video games designer in history. His impact on games can be compared to the influence of Shakespeare on theatre and Dickens on the novel. Like Dickens and Shakespeare, his work combines mainstream appeal with an unmatched level of inventiveness that places him in a different league to his peers. This is not to suggest that games are similar to plays or novels. A game isn't attempting to emulate the complex understanding of human nature that the best of those art forms achieve. It is an attempt to create a 'flow' state in the player. The player reacts to events on the screen, and the way in which they react alters those events. This creates a continuous feedback loop between the game and the gamer. Like so much else in the twentieth century, the link between the observed and the observer is fundamental.

To Miyamoto and his audience, of course, such concerns were unimportant. The distinction between highbrow and lowbrow was a meaningless excuse to look outside for validation. Postmodernism did not recognise the authority of any such external framework. Concepts such as highbrow and lowbrow, or 'art' and 'not-art', were projected on to the work by critics or gallery owners for their own benefit. They were not intrinsic qualities of the work itself. All that mattered, in games such as Super Mario Bros., was whether it was *in itself* any good.

The fact that a game such as Super Mario Bros. made total sense to an audience of children shows that the mainstream population was able to accept postmodernism, and take it in its stride, in a way that they never could with modernism.

Capitalism also had no problem with postmodernism. An example of this is the art world's response to the postmodern refusal to be either highbrow or lowbrow. This is nicely illustrated by their embrace of the American pop artist Roy Lichtenstein. Lichtenstein took frames from cheap comic books and copied them on to large canvases. Gallery owners were in no way concerned about Lichtenstein's copyright infringement. They took the view that his paintings were important art and that the comic-book images he blatantly plagiarised were not. A number of his paintings, such as *Sleeping Girl* from 1964 and 1961's *I Can See The Whole Room . . . And There's Nobody In It!*, have since sold for prices in excess of $40 million. To the business side of the art world, this is great stuff. The original comic art that those paintings were plagiarised from, meanwhile, is still viewed by gallery owners as either being essentially worthless, or as a curiosity that has become interesting due to the link to Lichtenstein. Comic-book artists are still not very happy about this.

Yet if audiences and the business establishment are so comfortable with postmodernism, why has it become such an undeniably hated movement? Trying to find someone who has anything good to say about postmodernism in the early twenty-first century is a challenge indeed. The word itself has become an insult, and one which negates the need to engage in further criticism. Once something has been dismissed as 'postmodern', it seems, it can be dismissed.

Postmodernism, as the word suggests, was what came after modernism. 'Modern' comes from the Latin word *modo*, meaning 'just now'. 'Post' meant 'after', so postmodernism essentially means 'after just now'. 'Modernism' may not have been a particularly helpful label for the avant garde culture of the early twentieth century, but it was positively descriptive compared to its successor.

It does not help that the term 'postmodern' has been applied so broadly. Eighties furniture that looked like it had been designed by designers on cocaine was postmodern. Comic books about characters who discovered that they were fictional were postmodern. Self-consciously awkward 1970s architecture was also postmodern. From Umberto Eco's *The Name of the Rose* to the pop videos of New Order, from the sculptures of Jeff Koons to the children's cartoon series *Danger Mouse*, postmodernism claimed them all. All this generated the suspicion that the term itself was meaningless. Many attempts to define it gave this impression too.

The reason for the current dismissal of postmodernism is its relationship with academia. The romance between academia and postmodernism, it is fair to say, did not end well.

Their relationship started promisingly enough. Postwar academia was quick to recognise postmodernism, and it had a lot to say about it. Many leading thinkers turned their attentions to the phenomenon and linked it to movements such as structuralism, post-structuralism and deconstruction. Postmodernism began to shape a great deal of intellectual debate, particularly in American academia. French philosophers such as Michel Foucault and Jacques Derrida became hugely influential. Yet as this process progressed, doubts started to appear. It wasn't apparent what use all this postmodern dialogue was, for a start. It didn't seem to produce anything solid. There was a nagging suspicion that it might be meaningless. Few people voiced that suspicion initially, for fear of looking ignorant, but increasingly it became hard to avoid the fact that a huge amount of academic postmodern discourse was gibberish.

This situation came to a head in 1996 when Alan Sokal, a physicist from New York University, submitted an article to the postmodernist academic journal *Social Text* entitled 'Transgressing the Boundaries: Towards a Transformative Hermeneutics of Quantum Gravity'. The article argued that reality was 'a social and linguistic construct' and that the development of a postmodern science would provide 'powerful intellectual support for the progressive political project'. Sokal was spoofing the deconstructionist idea that science was a socially constructed 'text', and hence open to different interpretations, by arguing that the laws of physics themselves could be anything we wanted them to be. He was making mischief, essentially, and his article was deliberately absurd and meaningless. But this was not apparent to the editorial team at *Social Text*, who thought that it was just the sort of thing that they were looking for and proceeded to publish it.

In normal circumstances Sokal's hoax would have been viewed as an attack on the world of academic publication. The failure of judgement of the journal's editors was, ironically, just the sort of thing that deconstructionists were banging on about when they talked about science being a social text. But thanks to the amount of unease surrounding postmodernism in academia, the Sokal hoax became viewed as a killer blow not to academic journals, but to postmodernism itself.

In the aftermath of the Sokal hoax, philosophers were very quick to leave postmodernism behind them, as can be seen in the string of critical obituaries that followed the death in 2004 of Jacques Derrida, the French founder of deconstructionism. The *New York Times* headline ran 'Jacques Derrida, Obtuse Theorist, Dies at 74'. It might have been thought that such an influential figure would have received a little more respect immediately after his death, but by then the world of philosophy was deeply ashamed about its postwar postmodern phase, and was distancing itself from the embarrassment as much as possible.

The problem was that there was no mechanism inside postmodernism for weeding out the meaningless from the meaningful. As a result it became possible to build an academic career by sounding clever, rather than being clever. Writing in *Nature* in 1998, the English biologist Richard Dawkins highlights the following example of apparently meaningless postmodern discourse: 'We can clearly see that there is no bi-univocal correspondence between linear signifying links or archi-writing, depending on the author, and this multireferential, multidimensional machinic catalysis. The symmetry of scale, the transversality, the pathic non-discursive character of their expansion: all these dimensions remove us from the logic of the excluded middle and reinforce us in our dismissal of the ontological binarism we criticised previously.' After a few decades of this sort of stuff, philosophers had had enough. It is understandable that anyone who had spent their working life reading texts like this would have rushed to put the boot into postmodernism, once Sokal had got it to the floor.

For academics, postmodernism was like quicksand. Once you fell into it, it was almost impossible to climb out. The more you struggled, the further in you were pulled. It also seemed inherently smug and pleased with itself. As an example, consider the way that this chapter used an old video game to explain postmodernism. This was, in itself, an extremely postmodern thing to do. It was an example of seemingly unrelated concepts thrown together and expected to work. It managed to avoid being either highbrow or lowbrow. This chapter has now started discussing itself, which shows that it is self-aware. This display of self-awareness essentially demonstrates the point that this paragraph was undertaken to explain, which makes it self-justifying, which in turn makes it even more postmodern and hence validates itself further. You can see why postmodernism winds people up.

Perhaps academia and postmodernism are fundamentally incompatible. Postmodernism denied that there was an external framework which could validate its works, yet that's exactly what academia was: a system to categorise and understand knowledge in relation to a rigid external framework. Postmodernism's rejection of external frameworks suggested that there were flaws in the foundation of academia. This embarrassed academia in the same way that Gödel's Incompleteness Theorem embarrassed logically minded mathematicians. In these circumstances, the speed with which the orthodoxy became an insult is understandable.

But outside of academia, postmodernism continued to spread through culture, entirely indifferent to the squabbles it was causing. One arena in which we can see its influence was religion and spirituality. As we've already noted, the spiritual model of subservience to a higher master, who protected and threatened punishment, had been undermined by individualism. The search for replacement models was under way and, in this atmosphere, those models could not help but be extremely postmodern.

During the 1960s and 1970s it was hard to avoid the idea that mankind was about to enter some form of glorious New Age. This was succinctly expressed in the opening lyrics of Broadway's attempt to capture the zeitgeist of the late 1960s, *Hair: The American Tribal Love-Rock Musical.* This opened with a song entitled 'Aquarius'. This celebrated the dawning of the Age of Aquarius, a reference to the fact that the astrological constellation behind the rising sun slowly changes over time. It takes 2,150 years to move from one constellation to the next. The twentieth century had begun towards the end of the Age of Pisces, and this astrological era was coming to an end. The Age of Pisces had coincided, rather neatly for a constellation with a fish symbol, with the Christian era. But the Age of Pisces would soon give way to the Age of Aquarius, and a number of people considered this to be spiritually significant. Carl Jung was one. He wrote that 1940 'is the year when we approach the meridian of the first star in Aquarius. It is the premonitory earthquake of the New Age.'

This was the idea that the cast of *Hair* were celebrating. Their opening song began with a number of astrological claims about the moon being in the seventh house and Jupiter aligning with Mars, which it claimed would usher in a period of love and peace. This sounded like a pretty good scenario. It was, unfortunately, largely meaningless. As the astrologer Neil Spencer has noted, Jupiter aligns with Mars every few months, and the moon is in the seventh house every day.

The opening lines to 'Aquarius', then, can tell us a lot about the New Age movement. It was hugely positive and creative, felt fresh and exciting, and was delivered with passion by wildly dressed, beautiful young people who would very shortly be getting naked. But what they were saying didn't hold up to scrutiny too well.

The New Age movement was not noted for its firm foundations and, like postmodernism itself, it is routinely mocked for this reason. Yet to dismiss a spiritual movement on the rational grounds of factual inaccuracy is, in many ways, to miss the point. Religions and spirituality are maps of our emotional territory, not our intellect. The Christian faith, for example, uses a crucifix as its key symbol. Crucifixion was one of the most awful forms of torture in the Iron Age world, and the icon of the cross represents unimaginable suffering. The sight of the cross is intended to generate an emotional understanding of that suffering, rather than an intellectual one. Just as a joke is valid if it is funny, even though it is not true, so spiritual symbols succeed if they have emotional or psychological value, regardless of the accuracy of stories that surround them. To look at the symbol of the crucifix and question whether the events it represents really happened does miss the point.

The appearance of the New Age movement illuminates how the great perspective shift of the early twentieth century affected our emotional selves. The fact that it happened at all is remarkable in itself, since unforced, wide-scale spiritual shifts affecting a sizeable proportion of the population are historically rare. The New Age was a rejection of hierarchical spirituality focused on the worship of a lord, and instead promoted the individual self to the role of spiritual authority. This produced almost exactly the same results in the spiritual world that postmodernism produced in the cultural world. Many varied and contradictory viewpoints were declared, leading to a highly personalised mishmash of world religions and spiritual practices. It welcomed astrology, Daoism, shamanism, tarot, yoga, angels, environmentalism, Kabbalah, the Human Potential Movement, ancient wisdom traditions and many more. Anything that shed light on the dual role of the practitioner as both a self-contained individual, and also part of the wider whole, was on the table. New Agers were free to take what they wanted for their personal practice, and to reject the rest. They became spiritual consumers in a marketplace of traditions. For those who kept faith in the existence of absolute certainty, it was all incredibly annoying.

The nature of many New Age traditions often required practitioners to keep a number of contradictory worldviews in play at any one time. This was evident in the Western adoption of T'ai Chi, a Chinese martial art that consists of sequences of slow, precise movements. T'ai Chi, as many Western studies show, does work. Daily practice brings about many mental and physical benefits, from reducing blood pressure to increasing flexibility, and it is especially beneficial for sufferers of depression, anxiety, osteoarthritis, ADHD and fibromyalgia.

The way T'ai Chi works is by training practitioners to manipulate *qi*, an ancient Chinese concept that refers to a living energy not unlike 'The Force' in *Star Wars*. Qi, Western scientists will tell you, does not exist. Yet it is not simply the case that T'ai Chi produces positive results despite the fact that its traditional explanation is mistaken. For students of T'ai Chi, qi is very real. They can physically feel it moving through them during practice. Indeed, students have to be aware of it, for it is not possible to perform the exercises properly without that awareness. Western practitioners, therefore, have to accept the contradiction that their health benefits arise from their manipulation of something that does not exist.

It is perhaps not surprising that the New Age movement was so riddled with contradictions. In a century when even Bertrand Russell's intellect was insufficient to define a system of logic and mathematics free from paradox, it is harsh to criticise those exploring emotional territory for intellectual flaws. New Agers were attempting to understand how we could be both separate individuals and part of a holistic whole at the same time. A level of paradox was integral to the project.

New Age movements may have had a tendency to appear lacking in what wider society would consider an acceptable level of bullshit detection, but they were also an accurate reflection of where the human race found itself spiritually in the late twentieth century.

With postmodernism so widely derided, it can be hard to remember why it came about. It was not simply some embarrassing mistake, as its detractors paint it, or a foolish wrong turn that we should learn from and then move on. Looking at the breadth of its influence on recent history, from Super Mario Bros. to the New Age, it is clear that there was something more fundamental to the phenomenon than the squabbles of academia might suggest.

At the start of the century, relativity triggered a paradigm shift in the physical sciences. Einstein acknowledged that the omphalos from which we orientated our understanding of the universe didn't exist, and that the notion that there was a 'centre of the universe' was absurd. The concept of one true perspective was replaced by a multitude of differing perspectives. Measurements were only valid relative to the observer.

Physicists are understandably twitchy about the theory of relativity being used as a justification for any form of cultural or moral relativity. They will stress that the lesson of Einstein is not that 'everything is relative', but that his work reconciled competing relative perspectives into non-relative, objective space-time. This is true, but it is also slightly disingenuous. As we noted earlier, the strange thing about what happened at the start of the twentieth century was that people in many fields, ranging across art, politics, music and science, all made a similar leap at roughly the same time. It was not the case that artists and thinkers attempted to take inspiration from Einstein and failed to understand him properly. A number of artists were clearly grappling with the relationship between the observer and the observed, or with reconciling multiple perspectives, before he published his General Theory. If you were feeling brave you could argue that Einstein was a modernist scientist, although to do so would annoy a lot of physicists.

As the century progressed scientists and mathematicians, such as Heisenberg, Freud, Gödel and Lorenz, reinforced the non-existence of the omphalos and instead stressed uncertainty, incompleteness and our lack of a self-contained, non-paradoxical system. At the same time artists and philosophers such as Picasso, Korzybski, Joyce and Leary continued their exploration of the human psyche, and came to much the same conclusion. The multiple-perspective models they pioneered made possible the individualism preached by Crowley, Rand, Thatcher and The Rolling Stones. We were all separate, and different, and considered our own perspectives to be personally valid. Our worldview could only be absolute if we could force everyone else to share it, and not even Hitler or Stalin could achieve that.

Our personal realities, then, were relative. We simply did not have anything absolute to orientate ourselves to. The closest thing the people of the northern hemisphere had to a fixed point in their lives was Polaris, the North Star, the only point in the heavens that remains fixed over the span of a human life. And even Polaris wobbles a little.

We might not like this. We might curse relativity, and crave an absolute position. But that doesn't change the fact that we do not have one.

Postmodernism was not some regrettable intellectual folly, but an accurate summation of what we had learnt. Even if it was intellectual quicksand from which escape did not appear possible. And which nobody seemed to like very much.

By the early twenty-first century the entire edifice of postmodernism had become routinely rejected. That, unfortunately, tended to include all the understanding that led up to it. Our current ideology stresses that *of course there is an absolute*. Of course there is truth. Richard Dawkins makes this argument when he says that 'We face an equal but much more sinister challenge from the left, in the shape of cultural relativism – the view that scientific truth is only one kind of truth and it is not to be especially privileged.' Or as Pope Benedict XVI said in his pre-conclave speech in 2005, 'Today, a particularly insidious obstacle to the task of education is the massive presence in our society and culture of that relativism which, recognising nothing as definitive, leaves as the ultimate criterion only the self with its desires. And under the semblance of freedom it becomes a prison for each one, for it separates people from one another, locking each person into his or her own ego.' As Martin Luther King put it, 'I'm here to say to you this morning that some things are right and some things are wrong. Eternally so, absolutely so. It's wrong to hate. It always has been wrong and it always will be wrong.' Or to quote the British philosopher Roger Scruton, 'In argument about moral problems, relativism is the first refuge of the scoundrel.' The existence of absolute truth has also been declared by neoliberalists and socialists, by terrorists and vigilantes, and by scientists and hippies. The belief in certainty is a broad church indeed.

All these people disagree on what form this absolutism takes, unfortunately. But they're pretty sure that it exists.

This faith in absolute certainty is not based on any evidence for the existence of certainty. It can sometimes appear to stem from a psychological need for certainty which afflicts many people, particularly older men. Cultural debate in the early twenty-first century has, as a result, descended into a War of the Certain. Different factions, all of whom agree about the existence of absolute truth, are shouting down anyone who has a different definition of that absolute truth.

Fortunately, true absolutism is rare. Most people, scientists and non-scientists alike, unconsciously adopt a position of multiple-model agnosticism. This recognises that we make sense of the world by using a number of different and sometimes contradictory models. A multiple-model agnostic would not say that all models are of equal value, because some models are more useful than others, and the usefulness of a model varies according to context. They would not concern themselves with infinite numbers of interpretations as that would be impractical, but they understand that there is never only one interpretation. Nor would they agree that something is not 'real' because our understanding of it is a cultural or linguistic construct. Things can still be real, even when our understanding of them is flawed. Multiple-model agnostics are, ultimately, pretty loose. They rarely take impractical, extreme positions, which may be why they do not do well on the editorial boards of academic postmodern journals.

Multiple-model agnosticism is an approach familiar to any scientist. Scientists do not possess a grand theory of everything, but they do have a number of competing and contradictory models, which are valid at certain scales and in certain circumstances. A good illustration of this point is the satellite navigation device in a car. The silicon chip inside it utilises our understanding of the quantum world; the GPS satellite it relies on to find its position was placed in orbit by Newtonian physics; and that satellite relies on Einstein's theory of relativity in order to be accurate. Even though the quantum, Newtonian and relativity models all contradict each other, the satnav still works.

Scientists generally don't lose too much sleep over this. A model is, by definition, a simplified incomplete version of what it describes. It may not be defendable in absolutist terms, but at least we can find our route home.

There is still, however, a tendency to frame these contradictory models as part of a hidden absolute, perhaps in order to avoid the whiff of postmodernism.

The absolutist approach to the contradictory nature of scientific models is to say that while all those models are indeed flawed, they will be superseded by a grand theory of everything, a wonder theory that does not contain any paradoxes and which makes sense of everything on every scale. The 2005 book about the quest for a Theory of Everything by the Canadian science journalist Dan Falk was called *Universe on a T-Shirt*, due to the belief that such a grand theory would be concise enough, like the equation $E=mc^2$, to be printed on a T-shirt.

To a multiple-model agnostic, this idea is a leap of faith. It is reminiscent of Einstein's mistaken belief that quantum uncertainty must be wrong, because he didn't like the thought of it. Of course if such a theory were found, multiple-model agnostics would be out celebrating with everyone else. But until that day comes it is not justifiable to assume that such a theory is out there, waiting. It is a call to an external ideal that is currently not supported by the data. A scientist who says that such a theory must exist is displaying the same ideological faith as someone who says God must exist. Both could be brilliant, but presently we should remain relativist enough to recognise them as unproven maybes.

In 1981 the American pop artist Andy Warhol began a series of paintings of the dollar sign. Warhol was a commercial artist from Pittsburgh who found fame in the 1960s with his gaudily coloured, mass-produced screen-prints of cultural icons. His most famous work, a series of prints of a Campbell's soup can, probably captured the essence of the postwar Golden Age better than any other work of visual art.

The dollar sign paintings Warhol began in the 1980s were not the last work he did. He died in 1987, and mass-produced typically Warholian canvases until the end. It is tempting, however, to see the dollar sign as the moment he finally ran out of ideas. With the exception of an increasing preoccupation with death, there is little in his 1980s work that was new. The dollar sign paintings seem in many ways to be the conclusion of his life's work.

They were big paintings, over two metres tall and just short of two metres wide. Each single dollar canvas could take up an entire wall in a gallery. The experience of walking into a white space, empty except for huge, bright dollar signs, is an uncomfortable one. Initially it is tempting to dismiss them as superficial, but there remains a lingering doubt that maybe Warhol had a point. Perhaps there really was nothing else he could do but paint the dollar sign as big and bright as he could. Perhaps the neoliberalists were correct and their dollar god was the only genuine power in the world. Maybe we had had a valid omphalos all along.

Money, it seemed in the 1980s, was the only thing solid enough for people to orientate themselves by. Individualism and *Do What Thou Wilt* had become the fundamental principle of living, so the power to achieve your desires became all-important. That power was most potently distilled in the form of money. It was money that allowed you to do what you wanted, and a lack of money that stopped you. It did not matter that a world where the dollar sign was the only true subject of worship was fundamentally grim. In a postmodern culture, all such judgement calls were subjective. Our artists, thinkers and scientists were free to offer up alternatives to the court of public opinion. The fact that they failed to do so seemed telling.

In 1992, the American political scientist Francis Fukuyama published his most influential book, *The End of History and the Last Man*. Fukuyama argued that, with the collapse of the Soviet Union, the neoliberal argument had won. Capitalism was the only option on the table and liberal democracies were the only justifiable form of state. Fukuyama claimed that we had reached the predetermined, final form of society, an argument that was essentially teleological and therefore religious in nature. He wrote in a deliberately prophetic, evangelical tone, proclaiming the 'Good News' of the eternal triumph of the capitalist paradise.

In this context, Warhol's paintings of dollar signs made complete sense. Was this, then, the endpoint of the twentieth century? Was this how our story was going to end?

Fukuyama was, fortunately, entirely wrong, as he would now be the first to admit. He split from the neoconservative movement over the US invasion of Iraq, a conflict he initially was in favour of, and voted for Barack Obama in 2008.

The individualism that had fuelled the neoliberal triumph was not the end point to humanity that people like Fukuyama or Margaret

Thatcher once believed it to be. It was a liminal period. The twentieth century had been the era after one major system had ended but before the next had begun. Like all liminal periods it was a wild time of violence, freedom and confusion, because in the period after the end of the old rules and before the start of the new, anything goes.

The coming era would keep the individual freedom that had been so celebrated in the twentieth century, but it would wed it to the last thing that the likes of The Rolling Stones wanted. Individual freedom was about to connect to the one thing it had always avoided. Freedom was about to meet consequence, and a new period of history was about to begin. In the research centres of Silicon Valley, a feedback loop was being built.

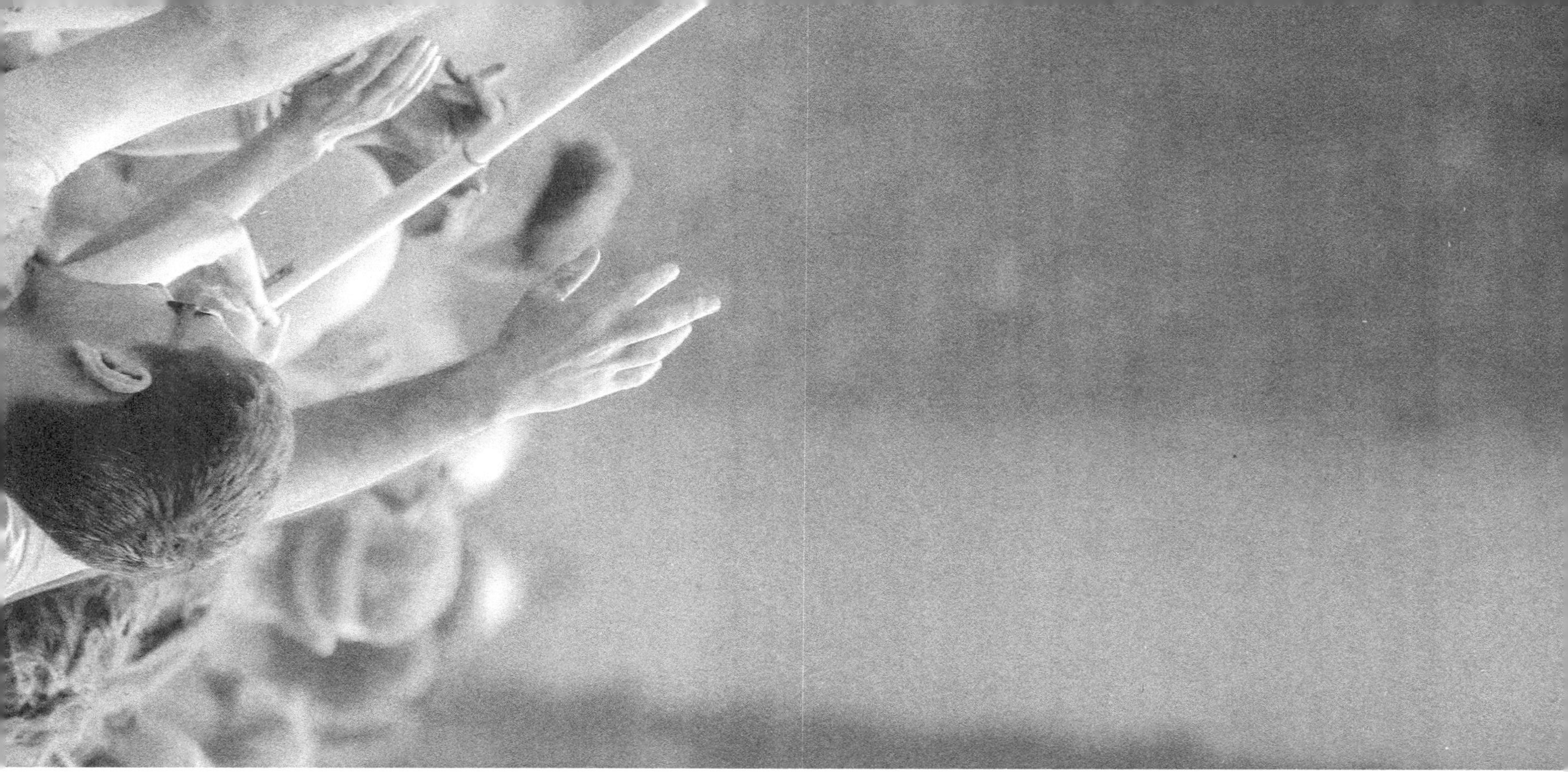

FIFTEEN: NETWORK

A Color Run participant in Barcelona taking a photo with a selfie stick, 2014 (Artur Debat/Getty)

A planet of individuals

The twenty-first century began over a twenty-four-hour period, which started at 11 a.m. on 31 December 1999. Or at least, that's how it appeared to those in Britain watching the BBC's coverage of the global New Year celebrations.

At 11 a.m. GMT it was midnight in New Zealand, which marked the occasion with a celebratory firework display in Auckland. This was beamed around the world by orbiting communication satellites, technology that was then about thirty-five years old and already taken for granted. Two hours later, a spectacular display over Sydney Harbour marked eastern Australia's entrance into the twenty-first century. The television coverage continued as the hours passed and a procession of major cities from East to West left the twentieth century behind.

In the Hebrew calendar the date was 22 Teveth 5760. Under the Islamic calendar it was 23 Ramadan 1420. To the UNIX computer operating system, it was 946598400. The significance of the moment was only a product of the perspective used to define it.

As we noted at the start, New Year's Day 2000 was not technically the beginning of the new millennium. That honour should have come a year later, on 1 January 2001. A few people were bothered by this, and wrote to newspapers, but were largely ignored. New Year's Eve 1999 was going to mark the end of the twentieth century, because that's how the planet of individuals wanted it. They had spent the last eighteen years listening to Prince singing that he wanted to 'party like it's 1999'. They were impatient to party that way, too. This was a world where the population drove cars and understood the strange appeal of watching all the digits in the milometer change at the same time. Who wanted to wait another year? Seeing 1999 turn into 2000 was clearly more fun than 2000 turn into 2001. People no longer recognised the claims to authority that institutions such as the Royal Observatory at Greenwich had once had. The twenty-first century began at midnight on 31 December 1999 by mutual consent. That's what people wanted.

In London, to the crowd enjoying a drunken singalong on the banks of the River Thames, the twenty-first century was a blank slate and full of potential. It still felt clean at that point, unsullied by what was to come. They had no inkling of 9/11, the War on Terror, or the coming global financial crash. Among the many songs that those revellers put their heart and soul into was a rendition of the classic Frank Sinatra standard 'My Way'. That song, perhaps more than any other, could symbolise the twentieth century. The lyrics are based around words such as 'I' or 'me', never 'we' or 'us'. *I'll* make it clear, he says, *I'll* state my case, of which *I'm* certain. It is far from the only song to base its lyrics on 'me, me, me', of course, but the proud manner of Sinatra's delivery marks it out as something special. When it was released it sounded heroic, a statement of pride in one man's ability to face life on his own terms. And yet, as we entered the twenty-first century, it was becoming possible to see that lyric in a different light. It is not the song of a man who understands himself to be part of something larger, or connected to a wider context in any meaningful way. It's the song of an isolated man who has spent his entire life attempting to force his own perspective on the world.

In the twentieth century, 'My Way' sounded glorious. It was a song frequently played at funerals to celebrate a life well lived. But by the twenty-first century, had it started to sound a little tragic?

The live media coverage of that event was not just a story of celebration and drunk people in fountains. Another story was unfolding at the same time, regarding something called the Millennium Bug. There was concern about how computer systems would deal with the date change.

When the earliest computers were built, computer memory was extremely expensive, so programmers employed neat little tricks to use the memory they had more efficiently. One such fudge involved not using the full four digits of a year, such as '1963', but instead just storing the last two digits, '63', and assuming that the year referred to the twentieth century. As computers developed over the 1970s and 1980s the cost of two storage bits became negligible, and tricks like this became unnecessary. But newer computing systems often had the legacy of older hardware and software in their DNA, and buried deep within the core code of many important systems was the assumption that after 1999 would come the year 1900.

Quite how significant this problem was was hard to tell. Computers had come from nowhere and, over the course of a couple of decades, they had taken over the running of pretty much everything. No one knew for sure exactly what the impact of the Millennium Bug would be. One potential scenario was doomsday. Planes would fall from the sky, the entire global financial system would cease to function, nuclear power stations would go into meltdown and mankind would be thrown back to the Stone Age. Another scenario was that nothing much would happen and the whole thing was a scam by programmers who wanted to up their overtime payments. Such a broad range of speculation was immensely attractive to the news media, who quickly rechristened the problem the 'Y2K Bug', on the grounds that it sounded more computery.

As a result of the concern, governments and companies spent a great deal of money updating their computer systems ahead of time, with some estimates placing the total cost as high as $600 million. When the year 2000 arrived without any significant problems there was great relief, and a nagging suspicion that the overtime-rich computer engineers had pulled a fast one.

What the Y2K Bug did do was force people to confront the extent to which they had become dependent on computers. The shift from a pre-digital to a post-digital society had been swift and total, and few people really noticed that it was happening until it was too late. There had been no time, therefore, to think through the most significant aspects of that revolution: all those computers were connected.

*

In the early twentieth century young children such as Wernher von Braun, Jack Parsons and Sergei Korolev had been shaped by the heroic, individualist science fiction of the age. Raised on Flash Gordon and Jules Verne, they dreamt of journeying into space and they dedicated their lives to realising that dream. The task was difficult, and only achieved through the catalyst of global war. The space rocket was born alongside its twin, the intercontinental nuclear missile. This had geopolitical consequences.

Before Hiroshima, the political and military game had been akin to chess. The king stayed at the back of the battlefield, as far away from the conflict as possible. To win the game it was necessary to take the opposing king, and all the other chess pieces were dedicated to making sure that this did not happen. Once the king was lost, the game was over. It was not necessary to completely wipe out your opponent in order to win, only to behead their master.

In the Cold War, different rules applied. The king could be taken immediately, during the first move of the game. It did not matter how far away from the frontline they hid because nuclear warheads were attached to rockets powerful enough to reach anywhere on the globe. It was just as easy to wipe out the White House or Red Square as it was to wipe out anywhere else. The game could be lost with entire armies still intact.

A rethink of the hierarchical power structure was called for. Previously, the king or tsar issued orders, which were passed down the chain of command. Their subjects carried out those orders and reported on their progress. That information was passed up the chain of command. Information could flow up or it could flow down, but it didn't do anything else.

In the new circumstances of the Cold War, the US military searched for a way to remain operative even if their command centre had been nuked and they were effectively headless. The answer was to ditch the hierarchical structure and design their information systems as a network. Every part of it should be able to contact every other part. Information needed to move from one part of the structure to another, and if the infrastructure along that route had been vaporised under a mushroom cloud, then that information should be able to find a different route to its destination.

In 1958, as a direct response to Sputnik, the US founded an agency within the US Department of Defense known as the Advanced Research Project Agency, or ARPA. In the 1960s ARPA began working on a computer network called ARPANET. It had not originally been designed specifically to remain operative in the event of a nuclear strike, but as the years progressed people came to think of it in those terms.

The ARPANET worked by parcelling up the information that had to flow between separate computers into standardised chunks known as 'packets'. Each packet was marked with its intended destination, but the route it took to get there was decided along the way and was dependent on the amount of traffic it encountered. A message from Los Angeles to San Francisco, for example, would be split into a number of packets that would set off across the network together, but they might not all take the same route to their destination. It was like a convoy of vehicles which got separated at busy traffic junctions, but which were still all able to make their own way to where they were going. When all the packets arrived in San Francisco they were recombined into the original message. When the speed of information flow was measured in milliseconds it did not matter if part of that data had come via the far side of the continent while the rest had taken a more direct route.

In the eyes of the ARPANET, geography was irrelevant. What mattered was that every node in the network was reachable from everywhere else. ARPA studied problems such as network efficiency and packets that got lost en route, and in time the ARPANET evolved into the internet we know today.

In the 1990s the internet reached beyond military and educational institutions and arrived in people's homes. A key factor in this was the 1995 release of Microsoft's Windows 95 operating system, which drove the growth of PC sales beyond the business world. Windows

95 was the first time that a software release was a major news story outside the specialist technology press. It arrived with great hype and a hugely expensive advertising campaign soundtracked by The Rolling Stones. Windows 95 was far friendlier than its predecessor, Windows 3.1, and did not require the same degree of specialist knowledge to use. It was also much better at running games.

What came next occurred so fast that most people didn't realise what had happened until years later. In the three years between the arrival of Windows 95 and the release of its successor, Windows 98, almost every aspect of the contemporary internet arrived in the mainstream. It may have been in embryonic form, but it was all there.

There was email, and a form of instant messaging known as IRC. There were webpages written in HTML, and web browsers with which to view them. There were communities to provide discussion forums, and there was music available in the form of mp3s if you knew where to look. There was a considerable amount of porn, as you might expect, and the first online shops. A few people began to put their diaries online, which seemed shocking in the context of twentieth-century ideas of privacy. These grew into the blogs we know today. Audio was first streamed in real time across the internet in 1995, thanks to software called RealAudio, and video followed in 1997. Interactive animation appeared online following the development of FutureSplash Animator, which later evolved into Adobe Flash.

To modern eyes, the internet of the 1990s would seem incredibly slow and unsophisticated. Yet at the point we entered the new millennium, all the concepts behind our current technology were already in place. We just didn't know what their impact would be.

The fact that everything became connected, potentially, to everything else, changed the way information flowed through our society. It was no longer the case that reports had to be passed upwards while orders were passed down. Every member of the network was free to do what they wished with the information that flowed their way. The result was an unexpected wave of transparency, which has washed over all our institutions and cast light on the secrets hidden within their structures.

In Britain, that process began in earnest with the MPs' expenses scandal. A culture of fraudulent expense claims had long been considered normal among British Members of Parliament, with MPs claiming taxpayers' money for everything from moat-clearing at country estates to a £1,645 house for ducks. This was exposed in 2009 and became a major scandal, with a number of politicians from both the Houses of Lords and Commons being either suspended, forced to resign, or prosecuted under criminal charges. Although much of the corruption was petty by historic standards, the scandal symbolised the growing distrust of authority and the public's desire for accountability. At the time, the story appeared to be simply good journalism by the *Daily Telegraph* newspaper, making use of the Freedom of Information Act. It soon proved to be the start of something far bigger.

The coming wave of transparency hit every institution in society. Corruption was uncovered in police forces. South Yorkshire Police, for example, were revealed to have altered numerous witness statements in a cover-up surrounding the deaths of ninety-six people at the Hillsborough Stadium in Sheffield in 1989, and to have fabricated evidence against ninety-five striking miners for political reasons in 1984. South Yorkshire Police looked like a model of probity compared to London's Metropolitan Police Force, who instructed undercover spies to 'smear' the family of the murdered teenager Stephen Lawrence and who regularly accepted bribes from journalists. The Fleet Street hacking scandal revealed the institutional level of corruption in British newspapers, most notably those owned by News International. Rebekah Brooks, the editor of the *Sun*, was charged with conspiracy to pervert the course of justice. She was eventually cleared, but it was still a shock to see such a powerful figure tried in court. Decades-old paedophile scandals suddenly came to light, most notably that of the late television personality Jimmy Savile. Savile's decades of child and sexual abuse prompted a major police investigation that revealed abuse by many well-known entertainment and establishment figures and the extent to which this was hushed up by institutions where the abuse took place, including the BBC and children's hospitals.

The public exposure of institutionalised corruption was not just a British phenomenon. The Catholic Church was revealed to be covering up child abuse within its ranks on a frankly unimaginable scale. Fraud was rampant in the banking world, as shown by the casual way bankers regarded the illegal rigging of an interest rate known as LIBOR for their own profits. There have been calls for transparency concerning everything, from corporate tax strategies, to the results of clinical trials, to the governance of international football.

In 2006 the website Wikileaks was founded. It revealed illegal activities in areas ranging from Peruvian oil to Swiss banking before causing a global political firestorm with the release of US war logs and diplomatic cables. The fact that no institution was safe from the sudden wave of transparency was evident when the former NSA contractor Edward Snowden released up to 1.7 million classified security documents, revealing the extent to which the NSA and GCHQ were operating without proper legal oversight and, essentially, spying on everyone. Snowden, like Wikileaks founder Julian Assange, was the product of a network culture in which the very fact that government secrets existed was reason enough to demand their exposure. Like the parliamentary expenses scandal, the establishment's desire to cover up the story was as damaging as the details that were leaked. By the time the constant outpouring of twenty-first-century institutional scandals had reached that most secretive of worlds, the security services, it was clear that something significant was occurring.

The reason for this wave of transparency was the arrival of the network. Previously, if an allegation had been made against Jimmy Savile, for example, that allegation was a lone piece of data which created cognitive dissonance with his fame and career. Even though there had always been rumours about Savile, a single allegation did not appear plausible to the eyes of the relevant authorities. It would not be properly investigated, not least because of his charity work and perceived power. But in the networked age, anyone who took an interest in such a lone piece of data could use the search algorithms of Google to discover other, previously isolated, pieces of data. When it was known that one child has made an allegation against Savile, it was easy to dismiss that as false. But when it was known that hundreds of different, unconnected children had made strikingly similar allegations, then the situation suddenly looked very different. Previously, victims did not come forward because they did not think that they would be believed. That they were willing to come forward in this network culture indicates that they recognised something had changed.

These investigations no longer needed to originate with professional journalists or regulatory bodies. Anyone with an interest or a grudge could start to collate all those separate claims. They could also connect to other concerned citizens who had stories to tell. Websites like the Everyday Sexism Project, which catalogues behaviour towards women on a day-to-day basis, showed how separate, often disregarded incidents were part of a serious, significant pattern.

People were no longer constrained by where they lived and who they knew. In the twentieth century a security analyst in Hawaii such as Edward Snowden would have had difficulty making contact with someone like Glenn Greenwald, the Brazil-based journalist noted for his work for the *Guardian* newspaper in London. But thanks to the internet, contacting him was trivial.

In the hierarchical world, cultures of corruption built up inside institutions because the constricted flow of information meant that those involved were safe from external observation. Now that information has been freed to flow in all directions, those cultures of corruption are exposed. In the twentieth century, President Nixon's link to the burglary at the Watergate Hotel was considered a major scandal. It required a huge amount of journalistic and legal process to deal with. Such a tiny scandal seems almost an irrelevance now, and the press and regulatory bodies are swamped by the mountains of institutionalised scandals they need to process.

The level of personal autonomy and freedom that the individualism of the twentieth century gave us is still with us in the twenty-first century. What has changed is that we can no longer expect to avoid the consequences of our actions. In August 2010, a forty-five-year-old woman in Coventry stopped to stroke a cat on a wall and then, in a moment of madness that she was unable to explain, picked the cat up and dropped it into a wheelie bin. The incident was caught on CCTV, spread virally round the internet, and the woman was very quickly identified. She was prosecuted, had to leave her job, and was subject to a global outpouring of anger and hatred which included Facebook pages calling her 'worse than Hitler'. Had the incident occurred before the network age, none of that would have happened. The woman would have walked away from the bin and continued with her life. No consequences would have resulted from her actions, with the exception of the consequences encountered by the cat.

The network has not just reorganised the flow of information around our society. It has imposed feedback loops into our culture. If what we do causes suffering, anger or repulsion, we will hear about it. Where once we regulated our behaviour out of fear of punishment by our Lord and master, now we adjust our actions in response to the buzzing cloud of verbal judgements from thousands of people. We are still free to choose our own path through society, in a way that we never were in the days of emperors, but we do have to take responsibility for our choices. This is bad news for libertarians who believe that there should be no limits placed on our freedom to place cats in bins, but it may prove to be good news for society as a whole.

If we are honest, it has been something of a shock to have what other people think revealed so publicly. Those raised in the twentieth century were perhaps unprepared for the amount of cynicism, tribal hatred and cruelty that you encounter every day on the internet. Many fear that the network itself is strengthening all this negativity, and that the echo chamber it provides is entrenching division. Yet at the same time, the more those parts of our psyches are placed in the light of transparency, the more we acknowledge, understand and recognise them for what they are. The young generation who grew up online can dispassionately avoid becoming sucked down into negativity through a shrug and an awareness that 'haters gonna hate'.

There are attempts being made to stop this process and shut down this flow of information. Organisations ranging from the Chinese Communist Party to Islamic states and American corporations have attempted to gain control of parts of the internet. These are, notably, all organisations with hierarchical structures. The original internet, free to access and neutral about the data it carries, might not survive long. In a similar way that the freedom and the lawlessness of the oceans became subject to international law and control in the eighteenth century, to the benefit of empires, so the internet as we know it may be replaced by a 'Balkanised' conglomeration of controlled networks.

It is not yet certain that this will be the future. Attempts to control the network are exposed by the transparency of the network itself. They also serve to reduce the legitimacy of the institutions attempting to gain control. Any attempts to disguise these actions and impose secrecy within an organisation affect that organisation's internal flow of information. This makes it less efficient, and therefore damages it. The wave of transparency will not be easily avoided.

Imagine that the people of this planet were points of light, like the stars in the night sky. Before the twentieth century we projected a constricting system onto these points, linking them in a hierarchical structure beneath a lord or emperor. This system informed our sense of identity, and governed how we orientated ourselves. It lasted for thousands of years. It may have been unfair and unjust, but it was stable.

At the start of the twentieth century that system shattered and those points were released to become free-floating, all with different perspectives. This was the postmodern relative world of individuals where, as Crowley put it, 'Every man and every woman is a star.'

Here was the twentieth century in all its chaotic glory, disjointed and discordant but wild and liberating.

But then a new system imposed itself on those free-floating points of light, just when we were least expecting it. Digital technology linked each of those points to, potentially, every other. Feedback loops were created, and consequences were felt. We could no longer be explained simply in terms of individuals. Now factors such as how connected we were, and how respected and influential we might be, were necessary to explain what was going on.

Money is, and has always been, important. But the idea that it was the *only* important thing was an oddity of the twentieth century. There had always been other social systems in place, such as chivalry, duty or honour, which could exert pressures that money alone could not. That has become the case again. Money is now just one factor that our skills and actions generate, along with connections, affection, influence and reputation.

Like the New Agers who saw themselves as individuals and, simultaneously, an integral part of a larger whole, we began to understand that what we were connected to was an important part of ourselves. It affected our ability to achieve our goals. A person that is connected to thousands of people can do things that a lone individual cannot. This generation can appear isolated as they walk the streets lost in the bubble created by their personal earphones. But they can organise into flashmobs in a way they never could before.

In the words of the American social physicist Alex Pentland, 'It is time that we dropped the fiction of individuals as the unit of rationality, and recognised that our rationality is largely determined by the surrounding social fabric. Instead of being actors in markets, we are collaborators in determining the public good.' Pentland and his team distributed smartphones loaded with tracking software to a number of communities in order to study the vast amount of data the daily interactions of large groups generated. They found that the overriding factor in a whole range of issues, from income to weight gain and voting intentions, was not individual free will but the influence of others. The most significant factor deciding whether you would eat a doughnut was not willpower or good intentions, but whether everyone else in the office took one. As Pentland discovered, 'The single biggest factor driving adoption of new behaviours was the behaviour of peers. Put another way, the effects of this implicit social learning were roughly the same size as the influence of your genes on your behaviour, or your IQ on your academic performance.'

A similar story is told by the research into child development and neuroscience. An infant is not born with language, logic and an understanding of how to behave in society. They are instead primed to acquire these skills from others. Studies of children who have been isolated from the age of about six months, such as those abandoned in the Romanian orphanages under the dictatorship of Nicolae Ceaușescu, show that they can never recover from the lost social interaction at that crucial age. We need others, it turns out, in order to develop to the point where we're able to convince ourselves that we don't need others.

Many aspects of our behaviour only make sense when we understand their social role. Laughter, for example, creates social bonding and strengthens ties within a group. Evolution did not make us make those strange noises for our own benefit. In light of this, it is interesting that there is so much humour on the internet.

Neuroscientists have come to view our sense of 'self', the idea that we are a single entity making rational decisions, as no more than a quirk of the mind. Brain-scanning experiments have shown that the mental processes that lead to an action, such as deciding to press a button, occur a significant period before the conscious brain believes it makes the decision to press the button. This does not indicate a rational individual exercising free will. It portrays the conscious mind as more of a spin doctor than a decision maker, rationalising the actions of the unconscious mind after the fact. As the Canadian-British psychologist Bruce Hood writes, 'Our brain creates the experience of our self as a model – a cohesive, integrated character – to make sense of the multitude of experiences that assault our senses throughout our lifetime.'

In biology an 'individual' is an increasingly complicated word to define. A human body, for example, contains ten times more non-human bacteria than it does human cells. Understanding the interaction between the two, from the immune system to the digestive organs, is necessary to understand how we work. This means that the only way to study a human is to study something more than that human.

Individualism trains us to think of ourselves as isolated, self-willed units. That description is not sufficient, either biologically, socially, psychologically, emotionally or culturally. This can be difficult to accept if you were raised in the twentieth century, particularly if your politics use the idea of a free individual as your primary touchstone. The promotion of individualism can become a core part of a person's identity, and something that must be defended. This is ironic, because where did that idea come from? Was it created by the person who defends their individualism? Does it belong to them? In truth, that idea was, like most ideas, just passing through.

In the late eighteenth century the English philosopher Jeremy Bentham designed a new type of prison called a panopticon. Through the design of this building and a complicated series of mirrors, a single watchman at the heart of the prison would be able to observe any prisoner he wished. Although this meant that at any one time almost all the prisoners were unobserved, each prisoner had no way of knowing if and when they were being watched. As a result, they had to act under the assumption that they were always being looked at. For Bentham, the power of the panopticon would come not through its efficiency, but from the effect constant potential observation would have on the consciousness of the prisoners.

This digital generation, born after 1990, have grown up in a form of communal panopticon. It has altered them in ways that their parents don't always appreciate. The older generation can view the craze for 'selfies', for example, as a form of narcissism. Yet those self-portraits are not just attempts to reinforce a personal concept of the individual self. They exist to be observed and, in doing so, to strengthen connections in the network. The culture of the 'selfie' may seem to be about twentieth-century individualism, but only when seen through twentieth-century eyes. Those photographs only become meaningful when shared.

In Sartre's 1943 philosophical book *Being and Nothingness* there is a short section entitled 'The Look'. Here Sartre attempts to deal with the philosophical nature of 'the other' by imagining an example which owes much to the multiple-perspective relativity of Einstein. Sartre talked about looking through a keyhole at another person, who is unaware that they are being observed, and about becoming completely absorbed in that observation. Suddenly the watcher realised that a third person had entered the room behind them, and that they themselves were being observed. Sartre's main point was concerned with the nature of objectification, but what is striking is how he described the awareness of being observed. For Sartre, being observed produces shame.

Compare this to the digital generation. Watch, for example, footage of the audience at a music festival, and note the reaction of the crowd when they suddenly realise that they are being looked at by television cameras. This is always a moment of delight and great joy. There is none of the shame that Sartre associated with being observed, and neither is that shame apparent in that generation's love of social media and seeming disregard for online privacy. Something has changed, therefore, in the sense of self which our culture instils in us, between Sartre's time and the present.

The millennial generation are now competing with the entire planet in order to gain the power that the attention of others grants. But they understand that the most effective way to get on in such an environment is to cooperate. This generation has intuitively internalised the lessons of game theory in a way that the people of the 1980s never did. They have a far greater understanding of consequence, and connections, than their grandparents. They understand the feedback loops that corporations are still not beholden to. It is no coincidence that when they organise, they do so in leaderless structures such as Occupy or Anonymous. They are so used to the idea that people come together to achieve a particular goal, and then disband, that most of what would technically be classed as their 'organisations' are never even formally named.

The great multinational companies that were built in the past were based upon new inventions or the control of a limited natural resource. More recent major corporations such as Google or Facebook are created by someone sitting down and writing them into existence. Programming is not about the manipulation of physical objects. It is about the manipulation of intangible things, such as information and instructions. For all the formality and structure inherent in its language and grammar, there have always been aspects of art and magic in coding. Reading and writing can pass information across space and time, but that information is essentially frozen and unable to do anything itself. Programming is like writing in a living language. It is text that acts on itself, and performs whatever tasks its author wishes it to undertake. Code never tires and is incredibly accurate. Language has become enchanted.

We have always been subject to change, but much of the nature and direction of that change has been beyond our control. Programming not only communicates, but acts on its own communication. As a consequence of the power this has given us, the speed of change has increased dramatically. But assuming the responsibility for steering change raises the issue of responsibility. It forces us to consider purpose: what are we trying to do? This was not a problem that blind chance or natural selection ever had to wrestle with.

In alchemy, there is a process known as *solve et coagula*. *Solve* means the process of reductionism or analysis; the reducing of a substance to its indivisible components. In the *I Ching* this is represented by hexagram 23, 'Splitting Apart'. This is equivalent to taking a pocket-watch to pieces in order to understand completely how it works, and is necessary before the equally vital second stage of the formula can commence. *Coagula* is the reassembly of the pocket-watch in a perfected, or at least improved, form. It is the equivalent of holism as opposed to reductionism, or the purposive process of synthesis as opposed to that of analysis.

In cultural terms, the individualism of the twentieth century can be likened to the process of *solve* taken to its logical end. Everything was isolated, and in isolation it was understood.

The network allows the process of *coagula*. Things are being reconnected, but with transparency and understanding that we did not have before the process of *solve*. Things are still liberated enough to be recombined into wild postmodern culture such as Super Mario Bros. But this can be done now with a sense of purpose and intention that was perhaps lacking before.

This does not mean, of course, that the hierarchies of the past or the individualism of the twentieth century are not still alive and kicking. Like all old ideas, they refuse to go gently and instead attempt to tear down what threatens them. Corporations are still machines for hoovering up wealth. Political power and financial power are still ignoring what the scientists of the world are screaming at them. Many non-Western states, particularly in the Islamic world, did not reject the hierarchical model and are finding the arrival of networks, without the buffer of a period of free individualism, to be painful. Organisations like Islamic State and Al Qaeda use the technology and structures of networks, but retain a hierarchical absolutist worldview. They have not undergone the difficult period of individualism which taught us how to deal with those who have different perspectives. The violence associated with absolutist views in a networked world, when seen in this light, seems almost inevitable.

The networked system is being tested, and it will only survive if it is strong. If this new system does make it, then who knows how long it will stand? Perhaps the network will last as our defining model for as long as the imperial system did. If that is the case, then free-floating, consequence-ignoring individualism was a brief liminal moment in history, a pause between breaths. The twentieth century will have been a rare time indeed, in those circumstances. It was a glimpse of mankind at its worst, and at its best. The supposed Chinese curse, 'May you live in interesting times', seems particularly apt.

For those who grew up in the twentieth century, things can look bleak. The fight over the freedom of the network looks futile. How is it possible to end growing inequality, and build a sustainable world, when the system continually makes the elite more powerful and everyone else weaker? Surely there is no way to protect ourselves from powerful institutions, from the NSA to Facebook, who are monitoring our lives and our personal identity for their own gain? But the millennial generation view hierarchical institutions differently. They do not automatically consider them legitimate. Legitimacy is something that needs to be justified in the networked world. It cannot be assumed. Such institutions are powerful, but without legitimacy they are not all-powerful. The Swiss corporation Nestlé was damaged more by the public's reaction to their promotion of formula milk in the developing world than they were by the legal response to their actions. The public's complaint would have been amplified greatly if it had occurred in the network age, rather than in the 1970s and 80s.

In the postmodern world, things made sense of themselves in isolation. In a network, things have context. Multiple perspectives are navigable, and practical. This is the age of realpolitik individualism. System behaviour is altered by changes of scale, as we have noted, and nowhere is that truer than with network growth.

The uncomfortable fact is that, if we take an honest look at what we know about climate science, the twenty-first century appears to be the penultimate century in terms of Western civilisation. That's certainly the position if we look at current trends and project forward. We can be sure, though, that there will be unpredictable events and discoveries ahead, and that might give us hope.

And there is the nature of the citizens of the twenty-first century to consider. If they were the same as the individuals of the twentieth century, then there would be little reason for hope. But they are not. As we can see from the bewildered way in which they shrug off the older generation's horror at the loss of privacy, the digital native generation do not see themselves using just the straitjacket of individualism. They know that model is too limited. They are more than isolated selves. Seeing themselves differently will cause them to act differently. Those of us born before the 1990s should, perhaps, get out of their way and wish them luck.

The network is a beheaded deity. It is a communion. There is no need for an omphalos any more.

Hold tight.

NOTES AND SOURCES

1 RELATIVITY: DELETING THE OMPHALOS

The description of the 1894 Greenwich Observatory bomb comes from the Royal Observatory website (*Propaganda by Deed: The Greenwich Observatory Bomb of 1894*, ROG Learning Team, www.rmg.co.uk), while the description of the weather comes from the Met Office archive of monthly weather reports (http://www.metoffice.gov.uk/archive/monthly-weather-report). The quote from Conrad comes from his novel *The Secret Agent*. William Gibson's comment about the future not being evenly distributed is a point he made in numerous interviews, for example his radio interview on *Talk of the Nation*, NPR, 30 November 1999.

The quote commonly attributed to Lord Kelvin is said to be from his lecture to the British Association for the Advancement of Science, 1900. The Albert Michelson quote is from *Light Waves and Their Uses*, p. 23. Philipp von Jolly's advice to Max Planck is quoted on p. 8 of Alan P. Lightman's *The Discoveries: Great Breakthroughs in Twentieth-Century Science*. H.G. Wells's June 1901 article 'Anticipations: An Experiment in Prophecy' was published in *North American Review* Vol. 172, No. 535, pp. 801–826.

My main source for details of Einstein's life was Walter Isaacson's wonderful biography *Einstein: His Life and Universe*. This is the source for Einstein's letter to Conrad Habicht (p. 21) and the quote from Chaim Weizmann at the end of the chapter (p. 282).

The discussion about relativity itself is the result of endless rereads of Einstein's own *Relativity, the Special and the General Theory: A Popular Exposition*. The different usage of Potsdamer Platz, Trafalgar Square and Times Square can be found in the original German edition (1917), the English translation (1920) and the Project Gutenberg eBook.

2 MODERNISM: THE SHOCK OF THE NEW

Much of the recent interest in Baroness Elsa von Freytag-Loringhoven is a result of a 2003 biography by the Canadian academic Irene Gammel and the 2011 publication of her letters and poems. The details of her life in this chapter are based on Gammel's biography, *Baroness Elsa*. The account of her performance for George Biddle appears on pp. 201–2 of that book, while her performance with a Duchamp press cutting is detailed on p. 173. The quote from Duchamp about the baroness is from Kenneth Rexroth's *American Poetry in the Twentieth Century*, p. 77.

Much of the discussion of Duchamp's work is inspired by the 2013 exhibition at the London Barbican, *The Bride and the Bachelors: Duchamp with Cage, Cunningham, Rauschenberg and Johns*. For more on the reputation of *Fountain*, see 'Duchamp's Urinal Tops Art Survey', BBC News, 1 December 2004. The quote from Jasper Johns is from pp. 109–10 of Pierre Cabanne's *Dialogues with Marcel Duchamp*. Duchamp spoke about the spectator's role in art in 'Session on the Creative Act', a 1957 talk at the Convention of the American Federation of Arts, Texas. Erik Levi spoke of 'the abyss of no tonal centre' in *The Sound and the Fury: A Century of Music*, BBC4, 12 February 2013. James Joyce explains how *Ulysses* would capture Dublin in its entirety in chapter four of Frank Bugden's *James Joyce and the Making of Ulysses*. The quote from Le Corbusier is from his 1925 book *Urbanisme*. Joyce's comments about giving up his immortality are taken from Richard Ellmann's *James Joyce*. Joyce's interview with Max Eastman in *Harper's* magazine is quoted in the same book.

Colin Wilson's description of peak awareness is on p. 171 of his book *Super Consciousness: The Quest for the Peak Experience*. The words of William Blake are taken from his 22 November 1802 letter to Thomas Butt, which is printed in *The Letters of William Blake* edited by Geoffrey Keynes. Dalí claimed to have been inspired by a camembert cheese on p. 317 of his book *The Secret Life of Salvador Dalí*. The quote from Friedrich Nietzsche is from his book *Human, All Too Human*.

3 WAR: HOIST THAT RAG

Details of Joshua Norton, including Greg Hill's quote, can be found in Adam Gorightly's *Historia Discordia*. For an anthropological explanation of how societies evolve into empires, see Jared Diamond's *Guns, Germs and Steel*, which also includes Diamond's quote about not killing strangers on p. 273.

The words of Joe Armstrong are from a recorded interview in the Imperial War Museum's *Voices of the First World War* series (http://www.1914.org/podcasts/podcast-8-over-by-christmas/). The quotes from Christopher Clark are taken from p. 562 and p. 561 of his book *Sleepwalkers*. The details of the assassination of the Archduke are based on Tim Butcher's *The Trigger*.

4 INDIVIDUALISM: DO WHAT THOU WILT

Aleister Crowley's descriptions of Aiwass and Aiwass's voice come from his book *Magick* (p. 427 and p. 435). The account of Crowley's time in Cairo is largely based on chapter four of Lawrence Sutin's *Do What Thou Wilt*. The three lines quoted from Crowley's *The Book of the Law* come from chapter III line 60, chapter I line 3 and chapter II line 23.

The quote from Dan Neil is taken from Tom Geoghegan's 1 July 2011 article for BBC News, 'Is the British roundabout conquering the US?'. The quotes from Ayn Rand's novella *Anthem* are from chapter eleven. Crowley's letter was written to Ethel Archer on 26 March 1947 and is quoted in Anthony Clayton's *Netherwood*. Information about the sales of *The Satanic Bible* comes from Chris Mathews's *Modern Satanism*. LeVey's quote that his work is just 'Ayn Rand, with trappings' is from Jesper Aagaard Petersen's *Contemporary Religious Satanism* (p. 2). Paul Ryan's speech is archived at http://www.prweb.com/releases/2012/4/prweb9457144.htm. For figures on the decline in church attendance in Europe, see http://viaintegra.wordpress.com/european-church-attendance. 'Love thy neighbour as thyself' is from the New Testament, Matthew 22:39.

The quote from Mussolini is found in *Rational Man* by Henry Babcock Veatch. Crowley's comments about free will are from *The Message of the Master Therion (Liber II)*. The quoted lines from his *Book of the Law* are 2:21 and 2:58. Details of the 2011 UK census come from the Office of National Statistics, and can be found online at http://www.ons.gov.uk/ons/guide-method/census/2011/index.html.

5 ID: UNDER THE PAVING STONES, THE BEACH

The Sergei Grigoriev quote is taken from Thomas Forrest Kelly's *First Nights: Five Musical Premieres* (p. 317), and the information regarding the missing police files comes from a 29 May 2013 BBC News Magazine report by Ivan Hewett, 'Did *The Rite of Spring* really spark a riot?' (http://www.bbc.co.uk/news/magazine-22691267).

Jean Cocteau's quote comes from his 1918 book, *Le Coq et l'Arlequin*, while Florent Schmitt's abuse is quoted on p. 82 of Alex Ross's *The Rest Is Noise*. The study of contemporary press reports mentioned is the 1971 PhD dissertation by Truman Campbell Bullard, *The first performance of Igor Stravinsky's Sacre du Printemps*. Leonard Bernstein's description of *The Rite of Spring* is taken from *The Rite of Spring at 100* (https://www.theriteofspringat100.org/the-history/). Stravinsky discussed the vision that inspired his composition in his 1936 autobiography. The quote from Sasha Waltz comes from a 27 May 2013 article in the *Guardian* by Kim Willsher, 'Rite that caused riots', and the conversation between Stravinsky and Diagev is taken from p. 81 of Alex Ross's *The Rest Is Noise*.

Sigmund Freud's famous quote that 'Dreams are the royal road to the subconscious' comes from his 1899 book *The Interpretation of Dreams*. Luis Buñuel's third-person comments are from his essay *Notes on the making of Un Chien Andalou*, which is included in the British Film Institute Blu-ray release of *L'Âge d'or*. The description of Dalí and Buñuel's dreams comes from Robert Short's essay *Un Chien Andalou and L'Âge d'or* in the same release. Buñuel's quote 'There's the film, let's go and make it' is from Meredith Etherington-Smith's biography of Dalí, *The Persistence of Memory* (p. 94). The review of *L'Âge d'or* in *Le Figaro* is from the 7 December 1930 issue. The details about the amount of fabric used in contemporary dresses comes from Bill Bryson's *One Summer*. The quote from Dorothy Dunbar Bromley is found in Joshua Zeitz's *Flapper*.

The quote from the Marquis de Sade is taken from Elaine Sciolino's 22 January 2013 *New York Times* article, 'It's a Sadistic Story, and France Wants It' (p. C1). Details of Dalí's sexuality can be found in Clifford Thurlow's *Sex, Surrealism, Dalí and Me*. His quote about impotence is found in Ian Gibson's *The Shameful Life of Salvador Dalí*. The praise from Freud appears in the foreword of Dalí's own autobiography, *The Secret Life of Salvador Dalí*. Henry Miller's description of Dalí is from an autograph which can be seen at http://www.openculture.com/2013/09/dali-is-the-biggest-prick-of-the-20th-century.html.

The number given for Jews killed in the Second World War is from Lucy

Dawidowicz's *The War against the Jews*. The detail about Hitler's portrait of Henry Ford is from p. 296 of Antony Beevor's *The Second World War*.

6 UNCERTAINTY: THE CAT IS BOTH ALIVE AND DEAD

Bertrand Russell's letter to his friend Helen Thomas is quoted on p. 179 of William R. Everdell's book *The First Moderns*. Einstein's quote about the ground being pulled out from underneath is found in Paul Arthur Schilpp's *Albert Einstein: Philosopher-Scientist, Volume II*. The quote from Richard Feynman is from *The Character of Physical Law*, a Cornell University lecture he gave in 1964, and the Douglas Adams reference is from *Mostly Harmless*. Einstein's quote that 'God doesn't play dice' can be found in many sources, for example p. 58 of William Hermanns's *Einstein and the Poet*. Stephen Hawking's essay *Does God Play Dice?* can be found on his website at http://www.hawking.org.uk/does-god-play-dice.html.

The quote 'a shlosh or two of sherry' is from Peter Byrne's December 2007 *Scientific American* article, 'The Many Worlds of Hugh Everett'. The quote from Léon Rosenfeld comes from a 2008 paper *The Origins of the Everettian Heresy*, by Stefano Osnaghi, Fabio Freitas and Freire Olival Jr, which is online at http://stefano.osnaghi.free.fr/Everett.pdf. The quote from David Deutsch is from his book *The Fabric of Reality*. The analogy about the size of atoms is from Marcus Chown's *Quantum Theory Cannot Hurt You*.

7 SCIENCE FICTION: A LONG TIME AGO IN A GALAXY FAR, FAR AWAY

The quotes from Alejandro Jodorowsky are from the 2013 documentary *Jodorowsky's Dune*, by Frank Pavich. The quote from Ken Campbell is as recounted to the author by Campbell's daughter Daisy Eris Campbell. The words of J.G. Ballard are taken from Alan Moore's essay 'Frankenstein's Cadillac', Dodgem Logic #4 June/July 2010, as are Moore's comments about Tom Swift. The quote from Carl Jung about UFOs is from his 1959 book *Flying Saucers*.

Gene Roddenberry's words are from *The Birth of a Timeless Legacy*, on the *Star Trek* Series One DVD (CBS DVD PHE 1021). The quote from Eric Hobsbawm is from p. 288 of *Fractured Times*. The account of the origins of cinema is based on Mark Cousins's *The Story of Film*. The

Corbett–Fitzsimmons Fight was directed by Enoch Rector in 1897. The quote from Joseph Campbell is taken from the Introduction to *The Hero with a Thousand Faces*. Philip Sandifer's analysis of The Hero's Journey is on his website, at http://www.philipsandifer.com/2011/12/pop-between-realities-home-in-time-for.html.

8 NIHILISM: I STICK MY NECK OUT FOR NOBODY

The quote from screenwriter Æneas MacKenzie comes from the commentary by historian Rudy Behlmer on the Blu-ray release of *Casablanca* (BDY79791). The three quoted lines from Alexander Trocchi's *Cain's Book* are from p. 47, p. 56 and pp. 29–30 respectively. *Waiting for Godot* was voted the most significant English-language play by a British National Theatre poll of 800 playwrights in 1999. Beckett originally wrote the play in French, but was eligible in this poll because he did the English-language translation himself. Vivian Mercier's review of that play was in the *Irish Times* on 18 February 1956.

The quote from Albert Camus is from his 1952 essay *Return to Tipasa*. Colin Wilson's criticism of *Endgame* is taken from his 2009 book *Super Consciousness*. The quote from William Blake comes from *The Mental Traveller*, circa 1803. The origins of the term 'Beat Generation' are detailed in *Brewer's Famous Quotations* by Nigel Rees. Gregory Corso's remarks are from the 1986 film *What Happened to Kerouac?* by directors Richard Lerner and Lewis MacAdams.

9 SPACE: WE CAME IN PEACE FOR ALL MANKIND

William Bainbridge's comment is found in George Pendle's *Strange Angel* (p. 15), and p. 14 of that book is the source for the quote from a 1931 textbook, *Astronomy* by Forest Ray Moulton (p. 296). John Carter's description of Jack Parsons's impact on the field of solid-fuel rocketry comes from p. 195 of his book *Sex and Rockets*. That book is also the source for the quote from Dr John Stewart (p. 47), the rumour about the sex tape with his mother's dog (p. 183), the descriptions of his advert for prospective tenants (p. 103) and his occult ceremonies (p. 84).

The detail about Nazi rocket scientists reaching a height of sixty miles comes from Deborah Cadbury's *Space Race* (p. 11). Parsons's poem 'Oriflamme' is quoted on p. 218 of George Pendle's *Strange Angel*. The quote from Crowley about Parsons and Hubbard comes from John Carter's *Sex and Rockets* (p. 150), and Hubbard's warning is from p. 177 of the same book.

Hitler's reaction to the A-4 rocket is described in Deborah Cadbury's *Space Race* (p. 5). Bob Holman describes his childhood memories of a V-2 bomb in his 8 September 2014 article in the *Guardian*, 'I saw the devastation of war 70 years ago. It was not glorious.' Von Braun's involvement in acquiring slave labour from Buchenwald is detailed in Deborah Cadbury's *Space Race* (p. 343). Estimates for the number of casualties at Hiroshima and Nagasaki are taken from the Yale Law School Avalon Project, http://avalon.law.yale.edu/20th_century/mp10.asp. The quote from Eisenhower comes from his autobiography *The White House Years* (pp. 312–3). Von Neumann's use of game theory to argue for an unprovoked nuclear strike on Russia is discussed in chapter twelve of Paul Strathern's *Dr Strangelove's Game*.

Sergei Korolev's story is told in Deborah Cadbury's *Space Race*, including the account of his witnessing a butterfly (p. 87), and the same book recounts NASA's reaction to news of Yuri Gagarin's successful flight (p. 246). Von Braun presented three episodes of the TV series *Walt Disney's Wonderful World of Colour: Man in Space* (1955), *Man and the Moon* (1955) and *Mars and Beyond* (1957). A complete transcription of President Kennedy's speech to Congress can be found at http://www.jfklink.com/ speeches/jfk/publicpapers/1961/jfk205_61.html.

10 SEX: NINETEEN SIXTY-THREE
(WHICH WAS RATHER LATE FOR ME)

Details of the life of Marie Stopes are from Ruth Hall's biography, *Marie Stopes*, with her mother's argument in favour of direct action taken from p. 54 and her father's letter quoted on p. 21. Stopes claimed to be ignorant of homosexuality and masturbation until the age of twenty-nine on p. 41 of her book *Sex and the Young*.

The source for the average size of families in 1911 and 2011 is the Office for National Statistics. The quote from the Lambeth Conference of Anglican Bishops is from the 1920 *Lambeth Conference Report*, Resolution 70. The words of Archbishop Hayes appeared in the 18 December 1921 issue of the *New York Times*, and were quoted on p. 162 of Ruth Hall's *Marie Stopes*. The letter from an unnamed railway worker is quoted on p. 257 of Hall's book.

Senator Smoot's words are from 'National Affairs: Decency Squabble' in the 31 March 1930 issue of *Time* magazine. The quote from Marie Stopes's trial is from p. 216 of Ruth Hall's biography. The Doctor Who story mentioned was *Doctor Who: The Romans*, broadcast on BBC1 in January 1965. The song 'Rape' appears on Peter Wyngarde's self-titled album, released by RCA Victor in 1970. Betty Friedan's quote regarding sexual liberation being a misnomer is from her September 1992 interview with *Playboy* magazine.

11 TEENAGERS: WOP-BOM-A-LOO-MOP-A-LOMP-BOM-BOM

'Tutti Frutti' was named number one in the *Top 100 Records That Changed the World* chart in the August 2007 edition of *MOJO* magazine. The quote from the 1956 *Encyclopaedia Britannica Yearbook* is taken from Ken Goffman and Dan Joy's *Counterculture through the Ages* (p. 225). Details of the FBI's investigation into 'Louie Louie', including the quote from LeRoy New, come from Alexis Petridis's article in the 23 January 2014 issue of the *Guardian*, '"Louie Louie": The Ultimate Rock Rebel Anthem'. Details of the original lyrics of 'Tutti Frutti' come from Charles White's *The Life and Times of Little Richard*. 'The Wagon' was a 1990 single by Dinosaur Jr, and opens their album *Green Mind*.

The account of Keith Richards's legal difficulties in Arkansas comes from his autobiography *Life*, as does his quote about The Beatles aiming 'I Wanna Be Your Man' at his band (p. 158) and his statement that 'We needed to do what we wanted to do' (p. 123). The William Rees-Mogg editorial appeared in the 1 July 1967 issue of *The Times*. Paul McCartney sang about the love you take and the love you make in 'The End', from the *Abbey Road* album. The speeches from Margaret Thatcher are archived at http://www.margaretthatcher.org/document/106689.

Remarks about a neurological difference between pre-pubescent and adolescent brains are based on Sarah-Jayne Blakemore and Suparna Choudhury's 'Development of the Adolescent Brain: Implications for Executive Function and Social Cognition', *Journal of Child Psychology and Psychiatry* 47:3/4 (2006), pp. 296–312. The argument that the counterculture fed the consumer culture it railed against is made by Joseph Heath and Andrew Potter in *The Rebel Sell: How the Counterculture Became Consumer Culture*. Kurt Cobain sang about how teenage angst had paid off well in 'Serve The Servants', the opening track on Nirvana's 1993 album

In Utero. Ken Goffman's quote comes from *Counterculture through the Ages* (p. xvi)

12 CHAOS: A BUTTERFLY FLAPS ITS WINGS IN TOKYO

Von Neumann's plans to control the weather are discussed in Paul Strathern's *Dr Strangelove's Game* (p. 303). Owen Paterson's worrying understanding of climate change was widely reported, for example in Rajeev Syal's 30 September 2013 article in the *Guardian*, 'Global warming can have a positive side, says Owen Paterson'.

Accounts of the work of Lorenz and Mandelbrot are based on James Gleick's *Chaos*. Lorenz's seminal paper 'Deterministic Nonperiodic Flow' was published in the *Journal of Atmospheric Sciences* (1963). The quote from Mother Teresa is taken from chapter one of James Lovelock's book *The Revenge of Gaia*. Carl Sagan's quote is from his book *Pale Blue Dot* (pp. xv–xvi). The quote from Alan Bean comes from the official website for his artwork (http://www.astronautcentral.com/BEAN/LTD/WayWayUp. html). The observation that the position of the astronaut on the journey home correlates to how spiritually affected they were was made by Andrew Smith in his book *Moon Dust: In Search of the Men who Fell to Earth*.

13 GROWTH: TODAY'S INVESTOR DOES NOT PROFIT
FROM YESTERDAY'S GROWTH

For an overview of the change in extinction levels in the twentieth century, including the estimate that puts them at between 100 and 1,000 times the background rate, see Howard Falcon-Lang's 11 May 2011 BBC article, 'Anthropocene: Have humans created a new geological age?' at http://www.bbc.co.uk/news/science-environment-13335683. The figure for the Gross World Product is based on the value of the US dollar in 1990, and is from J. Bradford DeLong's 1998 study *Estimating World GDP 1 Million BC – Present*. Figures for global energy consumption are taken from the 16 February 2012 article 'World Energy Consumption – Beyond 500 Exajoules', which is online at http://www.theoildrum.com/node/8936.

The statistics about the number of corporations using the Fourteenth Amendment are taken from the 2003 documentary *The Corporation*, directed by Mark Achbar and Jennifer Abbott. For a comparison of the economic size of nations compared to corporations, see the 4 December

2000 paper by Sarah Anderson and John Cavanagh of the Institute for Policy Studies, 'Top 200: The Rise of Corporate Global Power'. The US Justice Department's inability to prosecute HSBC was widely reported, see for example Matt Taibbi's article 'Gangster Bankers: Too Big to Jail', *Rolling Stone*, 14 February 2013. For a short overview of the Bhopal disaster, see Tony Law's 3 December 2008 article for *Wired*, 'Bhopal, "Worst Industrial Accident in History"'. For an account of the legal action by Nestlé against War On Want, see 'The Formula Flap', *TIME* magazine, 12 July 1976.

Life expectancy figures are from Kevin G. Kinsella's 'Changes in Life Expectancy 1900–1990', for the *American Journal of Clinical Nutrition* (1992). For an account of the prospects of the millennial generation, see Elliot Blair Smith's 21 December 2012 Bloomberg report, 'American Dream Fades for Generation Y Professionals'. For an account of falls in life expectancy, see Sabrina Tavernise, 'Life Spans Shrink for Least-Educated Whites in the U.S.', *New York Times*, 20 September 2012.

The fact that the combined wealth of the eighty richest people is the same as the combined wealth of the poorest 3.5 billion is from Oxfam's report *Wealth: Having It All and Wanting More*, which can be found online at http://www.oxfam.org/sites/www.oxfam.org/files/file_attachments/ib-wealth-having-all-wanting-more-190115-en.pdf. For figures about the size of the derivatives market, see 'Clear and Present Danger', the *Economist*, 12 April 2012. The quote from Warren Buffett comes from his *Berkshire Hathaway Annual Report*, 2002. For the link between oil prices and GDP, see Rebeca Jiménez-Rodriguez and Marcelo Sánchez, *Oil Price Shocks and Real GDP Growth: Empirical Evidence for Some OECD Countries*, European Central Bank 2004. For more on the ownership of Bolivian water, see Jim Shultz, 'The Politics of Water in Bolivia', *The Nation*, 14 February 2005.

For an example of follow-up studies to *Limits to Growth*, see C. Hall and J. Day, 'Revisiting the Limits to Growth after Peak Oil', *American Scientist*, 2009 or Graham Turner, *Is Global Collapse Imminent? An Updated Comparison of the Limits to Growth with Historical Data*, University of Melbourne, August 2014. For details of a study that argues inequality makes environmental problems worse, see Nafeez Ahmed's 14 March 2014 article in the *Guardian*, 'NASA-funded study: industrial civilisation headed for "irreversible collapse"?'. Margaret Thatcher's 1989 address to the UN General Assembly can be found at http://www.margaretthatcher. org/document/107817.

14 POSTMODERNISM: I HAPPEN TO HAVE MR MCLUHAN RIGHT HERE

Details about the origins of Mario are from David Sheff's *Game Over: Nintendo's Battle to Dominate an Industry*. IGN's Best Game of All Time poll can be found at http://uk.top100.ign.com/2005/001-010.html.

The *New York Times* obituary 'Jacques Derrida, Obtuse Theorist, Dies at 74' was written by Jonathan Kandell and published on 10 October 2004. Richard Dawkins's example of seemingly meaningless postmodern discourse is from 'Postmodernism Disrobed', *Nature* 394, 9 July 1998, where he is quoting the French psychoanalyst Félix Guattari. The quote from Carl Jung comes from a personal letter he wrote to Peter Baynes on 12 August 1940. Neil Spencer's comments about the lyrics to 'Aquarius' are from p. 124 of his book *True as the Stars Above*.

Richard Dawkins's quote regarding cultural relativity is from 'Richard Dawkins' Christmas Card List', the *Guardian*, 28 May 2007. Pope Benedict XVI's quote is from 'Address of His Holiness Benedict XVI to the Participants in the Ecclesial Diocesan Convention of Rome', 6 June 2005. The quote from Martin Luther King Jr is from 'Rediscovering Lost Values', a sermon delivered at Detroit's Second Baptist Church, 28 February 1954. Roger Scruton's quote is from p. 32 of his book *Modern Philosophy*.

15 NETWORK: A PLANET OF INDIVIDUALS

For a 'morning after' account of the non-impact of the Year 2000 Bug, see the BBC News story 'Y2K Bug Fails to Bite' (http://news.bbc.co.uk/1/hi/sci/tech/585013.stm). The Everyday Sexism Project is at everydaysexism.com. For an account of the consequences of putting a cat in a bin, see Patrick Barkham's 19 October 2010 article 'Cat Bin Woman Mary Bale Fined £250', in the *Guardian*.

The quotes from Alex Pentland come from his 5 April 2014 *New Scientist* article, 'The death of individuality'. The quote from Bruce Hood comes from his book *The Self Illusion*. For more on the amount of bacteria cells in a human body, see Melinda Wenner, 'Humans Carry More Bacterial Cells than Human Ones', in the 30 November 2007 issue of *Scientific American*.

BIBLIOGRAPHY

Abbott, Edwin Abbott, *Flatland: A Romance of Many Dimensions* (Seely & Co., 1884)

Adams, Douglas, *Mostly Harmless* (William Heinemann, 1992)

Azerrad, Michael, *Our Band Could Be Your Life: Scenes from the American Indie Underground, 1981–1991* (Little, Brown and Company, 2001)

Bayly, C.A., *The Birth of the Modern World 1780–1914* (Blackwell Publishing, 2004)

Beevor, Antony, *The Second World War* (Weidenfeld & Nicolson, 2012)

Bishop, Patrick, *Battle of Britain* (Quercus, 2009)

Bryson, Bill, *One Summer: America 1927* (Doubleday, 2013)

Bugden, Frank, *James Joyce and the Making of Ulysses* (Oxford University Press, 1934)

Butcher, Tim, *The Trigger: Hunting the Assassin Who Brought the World to War* (Chatto & Windus, 2014)

Cabanne, Pierre, *Dialogues with Marcel Duchamp* (Da Capo, 1988)

Cadbury, Deborah, *Space Race: The Battle to Rule the Heavens* (Fourth Estate, 2005)

Campbell, Joseph, *The Hero with a Thousand Faces* (Pantheon, 1949)

Carter, John, *Sex and Rockets: The Occult World of Jack Parsons* (Feral House, 2004)

Chan, Stephen, *The End of Certainty: Towards a New Internationalism* (Zed Books, 2010)

Chown, Marcus, *Quantum Theory Cannot Hurt You: A Guide to the Universe* (Faber & Faber, 2007)

Clark, Christopher, *The Sleepwalkers: How Europe Went to War in 1914* (Penguin, 2012)

Clayton, Anthony, *Netherwood: Last Resort of Aleister Crowley* (Accumulator Press, 2012)

Cocteau, Jean, *Le Coq et l'Arlequin* (Éditions de la Sirène, 1918)

Conrad, Joseph, *The Secret Agent* (Methuen & Co., 1907)

Le Corbusier (Charles-Édouard Jeanneret-Gris), *Urbanisme* (Éditions Flammarion, 1925)

Cousins, Mark, *The Story of Film* (Pavilion, 2011)

Crowley, Aleister, *The Book of the Law* (Ordo Templi Orientis, 1938)

Crowley, Aleister, *Magick: Liber ABA, Book 4* (Red Wheel/Weiser, 1998)

Dalí, Salvador, *The Secret Life of Salvador Dalí* (Dial Press, 1942)

Davis, Stephen, *Hammer of the Gods: Led Zeppelin Unauthorised* (Sidgwick & Jackson, 1985)

Dawidowicz, Lucy, *The War against the Jews* (Holt, Rinehart and Winston, 1975)

Dawkins, Richard, *The Selfish Gene* (Oxford University Press, 1976)

Debord, Guy, *Society of the Spectacle* (Black & Red, 1970)

Deutsch, David, *The Fabric of Reality: The Science of Parallel Universes and Its Implications* (Penguin, 1998)

Diamond, Jared, *Guns, Germs and Steel: The Fates of Human Societies* (W.W. Norton, 1997)

Douds, Stephen, *The Belfast Blitz: The People's Story* (Blackstaff Press, 2011)

Einstein, Albert, *Relativity, the Special and the General Theory: A Popular Exposition* (Methuen & Co., 1920)

Eisenhower, Dwight D., *Mandate For Change, 1953–1956: The White House Years* (Doubleday, 1963)

Ellmann, Richard, *James Joyce* (Oxford Paperbacks, 1984)

Etherington-Smith, Meredith, *The Persistence of Memory: A Biography of Dalí* (Da Capo, 1995)

Everdell, William R., *The First Moderns: Profiles in the Origins of Twentieth-Century Thought* (University of Chicago, 1997)

Fitzgerald, F. Scott, *The Great Gatsby* (Charles Scribner's Sons, 1925)

Freud, Sigmund, *The Interpretation of Dreams* (Macmillan, 1899)

Friedan, Betty, *The Feminine Mystique* (W.W. Norton, 1963)

Gammel, Irene, *Baroness Elsa: Gender, Dada and Everyday Modernity* (MIT Press, 2003)

Gibson, Ian, *The Shameful Life of Salvador Dalí* (Norton & Co., 1997)

Gleick, James, *Chaos: Making a New Science* (Sphere, 1987)

Goffman, Ken and Joy, Dan, *Counterculture through the Ages: From Abraham to Acid House* (Villard, 2005)

Gorightly, Adam (editor), *Historia Discordia: The Origins of the Discordian Society* (RVP Press, 2014)

Graeber, David, *Debt: The First 5000 years* (Melville House, 2011)

Greer, Germaine, *The Female Eunuch* (Harper Perennial, 1970)

Hall, Ruth, *Marie Stopes: A Biography* (André Deutsch, 1977)

Heath, Joseph and Potter, Andrew, *The Rebel Sell: How the Counterculture Became Consumer Culture* (Capstone, 2005)

Heller, Joseph, *Catch 22* (Jonathan Cape, 1962)

Hermanns, William, *Einstein and the Poet: In Search of the Cosmic Man* (Branden Publishing, 1983)

Higgs, John, *I Have America Surrounded: The Life of Timothy Leary* (The Friday Project, 2006)

Hobsbawm, Eric, *The Age of Empire: 1875–1914* (Abacus, 1989)

Hobsbawm, Eric, *The Age of Extremes: 1914–1991* (Abacus, 1995)

Hobsbawm, Eric, *Fractured Times: Culture and Society in the Twentieth Century* (Little, Brown, 2013)

Hood, Bruce, *The Self Illusion: Why There Is No 'You' Inside Your Head* (Constable, 2013)

Huxley, Aldous, *Brave New World* (Chatto & Windus, 1932)

Isaacson, Walter, *Einstein: His Life and Universe* (Thorndike Press, 2007)

Johnson, Paul, *Modern Times: The World from the Twenties to the Nineties* (HarperCollins, 1991)

Joyce, James, *Ulysses* (Odyssey Press, 1932)

Judt, Tony with Snyder, Timothy, *Thinking the Twentieth Century* (Vintage, 2013)

Jung, C.G., *Flying Saucers: A Modern Myth of Things Seen in the Skies* (Princeton University Press, 1959)

Jung, C.G., *Memories, Dreams, Reflections* (Fontana, 1967)

Jung, C.G., *Synchronicity: An Acausal Connecting Principle* (Princeton University Press, 1973)

Kelly, Thomas Forrest, *First Nights: Five Musical Premieres* (Yale University Press, 2001)

Keynes, Geoffrey (editor), *The Letters of William Blake* (Macmillan, 1956)

Korzybski, Alfred, *Science and Sanity: An Introduction to Non-Aristotelian Systems and General Semantics*, second edition (Institute of General Semantics, 2010)

Kripal, Jeffrey J., *Esalen: America and the Religion of No Religion* (University of Chicago Press, 2007)

Kumar, Manjit, *Quantum: Einstein, Bohr and the Great Debate over the Nature of Reality* (Icon Books, 2009)

Lachman, Gary Valentine, *Turn Off Your Mind: The Mystic Sixties and the Dark Side of the Age of Aquarius* (Sidgwick & Jackson, 2001)

Lawrence, D.H., *Lady Chatterley's Lover* (Tipografia Giuntina, 1928)

Leary, Timothy, *Confessions of a Hope Fiend* (Bantam, 1973)

Lightman, Alan P., *The Discoveries: Great Breakthroughs in Twentieth-Century Science* (Random House, 2005)

Lovelock, James, *The Revenge of Gaia* (Penguin, 2006)

Marcus, Greil, *Lipstick Traces: A Secret History of the Twentieth Century* (Faber & Faber, 2001)

Markoff, John, *What the Dormouse Said: How the 60s Counterculture Shaped the Personal Computer Industry* (Viking, 2005)

Marr, Andrew, *A History of Modern Britain* (Macmillan, 2007)

Marr, Andrew, *The Making of Modern Britain: From Queen Victoria to VE Day* (Macmillan, 2009)

Mathews, Chris, *Modern Satanism: Anatomy of a Radical Subculture* (Greenwood Publishing Group, 2009)

McLuhan, Marshall, and Fiore, Quentin, *The Medium is the Massage: An Inventory of Effects* (Hardwired, 1967)

Michelson, Albert, *Light Waves and Their Uses* (University of Chicago, 1903)

Moulton, Forest Ray, *Astronomy* (Macmillan, 1931)

Nietzsche, Friedrich, *Human, All Too Human* (Cambridge University Press, 1878)

Packer, George, *The Unwinding: An Inner History of the New America* (Faber & Faber, 2013)

Pendle, George, *Strange Angel: The Otherworldly Life of Rocket Scientist John Whiteside Parsons* (Weidenfeld & Nicolson, 2005)

Petersen, Jesper Aagaard, *Contemporary Religious Satanism: A Critical Anthology* (Ashgate, 2009)

Rand, Ayn, *Anthem* (Cassell, 1938)

Rees, Nigel, *Brewer's Famous Quotations: 5000 Quotations and the Stories Behind Them* (Chambers, 2009)

Rexroth, Kenneth, *American Poetry in the Twentieth Century* (Herder and Herder, 1973)

Richards, Keith, with Fox, James, *Life* (Weidenfeld & Nicolson, 2010)

Roberts, Andy, *Albion Dreaming: A Popular History of LSD in Britain* (Marshall Cavendish, 2012)

Ross, Alex, *The Rest Is Noise: Listening to the Twentieth Century* (Fourth Estate, 2007)

Rushkoff, Douglas, *Program or be Programmed: Ten Commandments for a Digital Age* (Soft Skull Press, 2010)

Sagan, Carl, *Pale Blue Dot: A Vision of the Human Future in Space* (Ballantine, 1994)

Schilpp, Paul Arthur (editor), *Albert Einstein: Philosopher-Scientist, Volume II* (Harper, 1951)

Scruton, Roger, *Modern Philosophy: An Introduction and Survey* (Penguin, 1996)

Sheff, David, *Game Over: Nintendo's Battle to Dominate an Industry* (Random House, 1993)

Smith, Andrew, *Moon Dust: In Search of the Men who Fell to Earth* (Bloomsbury, 2009)

Spencer, Neil, *True as the Stars Above: Adventures in Modern Astrology* (Orion, 2000)

Stopes, Marie C., *Sex and the Young* (Putnam, 1926)

Strathern, Paul, *Dr Strangelove's Game: A Brief History of Economic Genius* (Penguin, 2002)

Strausbaugh, John, *E: Reflections of the Birth of the Elvis Faith* (Blast Books, 1995)

Stravinsky, Igor, *An Autobiography* (W.W. Norton, 1936)

Sutin, Lawrence: *Do What Thou Wilt: A Life of Aleister Crowley* (St Martin's Griffin, 2000)

Thompson, Hunter S., *Fear and Loathing on the Campaign Trail '72* (Flamingo, 1973)

Thurlow, Clifford, *Sex, Surrealism, Dalí and Me: The Memoirs of Carlos Lozano* (Razor Books, 2000)

Trocchi, Alexander, *Cain's Book* (Grove Atlantic, 1960)

Vallée, Jacques, *The Network Revolution: Confessions of a Computer Scientist* (And/Or Press, 1982)

Veatch, Henry Babcock, *Rational Man: A Modern Interpretation of Aristotelian Ethics* (Indiana University Press, 1962)

Warncke, Carsten-Peter, *Picasso* (Border Press, 1998)

Watson, Peter, *A Terrible Beauty: The People and Ideas that Shaped the Modern World* (Weidenfeld & Nicolson, 2000)

White, Charles, *The Life and Times of Little Richard: The Quasar of Rock* (Da Capo, 1984)

Wilson, Colin, *Super Consciousness: The Quest for the Peak Experience* (Watkins, 2009)

Wilson, Robert Anton and Shea, Robert, *The Illuminatus! Trilogy* (Dell, 1975)

Woolf, Virginia, *A Room of One's Own* (Hogarth Press, 1929)

Yeats, W.B., *The Complete Poems of W.B. Yeats* (Wordsworth Editions, 2000)

Zeitz, Joshua, *Flapper: A Madcap Story of Sex, Style, Celebrity, and the Women who made America Modern* (Three Rivers Press, 2007)

ACKNOWLEDGEMENTS

Huge thanks for the invaluable help from, in no sensible order: Joanne Mallon, Jason Arnopp, CJ Stone, Steve Moore, Alan Moore, Alistair Fruish, Bea Hemming, Zoe Ross, Shardcore, Sarah Ballard, Jenny Bradshaw, Zoë Pagnamenta, Nathalie Grindles, Linda Shaughnessy, Holly Harley, Zenbullets, Alice Whitwham, Margaret Halton, Amy Mitchell, Joanne Hornsby, Leanne Oliver, Ilaria Tarasconi, John Marchant, Jamie Reid and Mark Pilkington.

INDEX

New Deal, 226; and rehabilitation of Nazis, 176–7; McCarthyism, 139, 182; invasion of Iraq, 291; corporate influence in, 263–6; government spending, 265; increasing inequality, 260–1; life expectancy, 261
US Constitution, Fourteenth Amendment, 252–4
universal suffrage, 69, 210
universe, 167–8, 233; centre of, 24, 167, 286

V-2 rockets, 174, 179, 182–3, 189
Valens, Ritchie, 217
Verne, Jules, 133, 165, 168–9, 300
Verve, The, 219
Victoria, Queen, 57
Vietnam War, 223
Volcker, Paul, 264

Wall Street Crash, 81
Wallace, Mike, 78
Wallis, Hal, 149
Waltz, Sasha, 93
Warhol, Andy, 290–1
Watergate scandal, 305
Watts, Alan, 221
Webern, Anton, 43
Weimar Republic, 182, 258
Weir, Bob, 207

Weizmann, Chaim, 31
Welles, Orson, 130
Wells, H.G., 19, 133
Whitehead, Alfred North, 109
Whore of Babylon, 169, 171–2
Wiccan faith, 246
Wikileaks, 304
Wild Angels, The, 218
Wild One, The, 218
Wilhelm II, Kaiser, 57, 59
Wilhelm, Richard, 158
Wilson, Charles Erwin, 258–9
Wilson, Colin, 49, 154–5
Wilson, Woodrow, 56
Wire, The, 142
women's suffrage, 195–6, 202, 210
Woolf, Virginia, 201–2
Woolsey, Judge John, 47
World Bank, 265
World of Warcraft, 142
Worsley, Lucy, 4
Wren, Sir Christopher, 15
Wyngarde, Peter, 208

Yeats, W.B., 73; The Second Coming, 94

Zurich, 20, 25